Pelican Books

How to Lie with Statistics

Darrell Huff was born in 1913 in Iowa, and grew up there and in California. He received his B.A. ('With Distinction' and election to Phi Beta Kappa) and M.A. degrees from the State University of Iowa, where he did additional graduate work in social psychology, including work in statistics and mental testing. He has been associate or managing editor of several magazines, such as *Look* and *Better Homes & Gardens*, but for nearly twenty years he has been a free-lance writer of articles and occasional short stories for many magazines, among them *Harper's*, *Saturday Evening Post*, *Esquire* and the *New York Times Magazine*. He and his wife, also a writer, have lived in Spain, Mallorca, Italy, France, Greece, Germany, Denmark and in the United States.

Much of Darrell Huff's writing has to do with mathematics and his book *How to Take a Chance* is also published in Penguins. In 1963 he was awarded a National School Bell Award for his work.

With
pictures by
Mel Calman

How to Lie with Statistics

Darrell Huff

Penguin Books

Penguin Books Ltd, Harmondsworth,
Middlesex, England
Penguin Books, 625 Madison Avenue, New York,
New York 10022, U.S.A.
Penguin Books Australia Ltd, Ringwood,
Victoria, Australia
Penguin Books Canada Ltd, 2801 John Street,
Markham, Ontario, Canada L3R 1B4
Penguin Books (N.Z.) Ltd, 182–190 Wairau Road,
Auckland 10, New Zealand

First published by Victor Gollancz 1954
Published in Pelican Books 1973
Reprinted 1974, 1975 (twice), 1976, 1977, 1978, 1979, 1980

Set, printed and bound in Great Britain by
Cox & Wyman Ltd,
Reading
Set in Linotype Pilgrim

To my wife

with good reason

There are three kinds of lies: lies, damned lies, and statistics.
– *Disraeli*

Statistical thinking will one day be as necessary for efficient citizenship as the ability to read and write.
– *H. G. Wells*

It ain't so much the things we don't know that get us in trouble. It's the things we know that ain't so.
– *Artemus Ward*

Round numbers are always false.
– *Samuel Johnson*

I have a great subject [statistics] to write upon, but feel keenly my literary incapacity to make it easily intelligible without sacrificing accuracy and thoroughness.
– *Sir Francis Galton*

Contents

Acknowledgements

The pretty little instances of bumbling and chicanery with which this book is peppered have been gathered widely and not without assistance. Following an appeal of mine through the American Statistical Association, a number of professional statisticians – who, believe me, deplore the misuse of statistics as heartily as anyone alive – sent me items from their own collections. These people, I guess, will be just as glad to remain nameless here. I found valuable specimens in a number of books too, primarily these: *Business Statistics*, by Martin A. Brumbaugh and Lester S. Kellogg; *Gauging Public Opinion*, by Hadley Cantril; *Graphic Presentation*, by Willard Cope Brinton; *Practical Business Statistics*, by Frederick E. Croxton and Dudley J. Cowden; *Basic Statistics*, by George Simpson and Fritz Kafka; and *Elementary Statistical Methods*, by Helen M. Walker.

Introduction

With prospects of an end to the hallowed old British measures of inches and feet and pounds, the Gallup poll people wondered how well known its metric alternative might be. They asked in the usual way, and learned that even among men and women who had been to a university 33 per cent had never heard of the metric system.

Then a Sunday newspaper conducted a poll of its own – and announced that 98 per cent of its readers knew about the metric system. This, the newspaper boasted, showed 'how much more knowledgeable' its readers were than people generally.

How can two polls differ so remarkably?

Gallup interviewers had chosen, and talked to, a carefully selected cross-section of the public. The newspaper had naïvely, and economically, relied upon coupons clipped, filled in, and mailed in by readers.

It isn't hard to guess that most of those readers who were unaware of the metric system had little interest in it or the coupon; and they selected themselves out of the poll by not bothering to clip and participate. This self-selection produced, in statistical terms, a biased or unrepresentative sample of just the sort that has led, over the years, to an enormous number of misleading conclusions.

A few winters ago a dozen investigators independently reported figures on antihistamine pills. Each showed that a considerable percentage of colds cleared up after treatment. A great fuss ensued, at least in the advertisements, and a medical-product boom was on. It was based on an eternally springing hope and also on a curious refusal to look past the statistics to a fact that has been known for a long time. As Henry G. Felsen, a humorist and no medical authority, pointed out quite a while ago, proper treatment will cure a cold in seven days, but left to itself a cold will hang on for a week.

So it is with much that you read and hear. Averages and relationships and trends and graphs are not always what they seem. There may be more in them than meets the eye, and there may be a good deal less.

The secret language of statistics, so appealing in a fact-minded culture, is employed to sensationalize, inflate, confuse, and oversimplify. Statistical methods and statistical terms are necessary in reporting the mass data of social and economic trends, business conditions, 'opinion' polls, the census. But without writers who use the words with honesty and understanding and readers who know what they mean, the result can only be semantic nonsense.

In popular writing on scientific matters the abused statistic is almost crowding out the picture of the white-jacketed hero labouring overtime without time-and-a-half in an ill-lit laboratory. Like the 'little dash of powder, little pot of paint', statistics are making many an important fact 'look like what she ain't'. A well-wrapped statistic is better than Hitler's 'big lie'; it misleads, yet it cannot be pinned on you.

This book is a sort of primer in ways to use statistics to deceive. It may seem altogether too much like a manual for swindlers. Perhaps I can justify it in the manner of the retired burglar whose published reminiscences amounted to a graduate course in how to pick a lock and muffle a footfall: the crooks already know these tricks; honest men must learn them in self-defence.

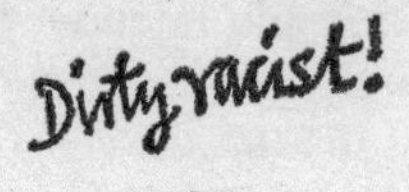

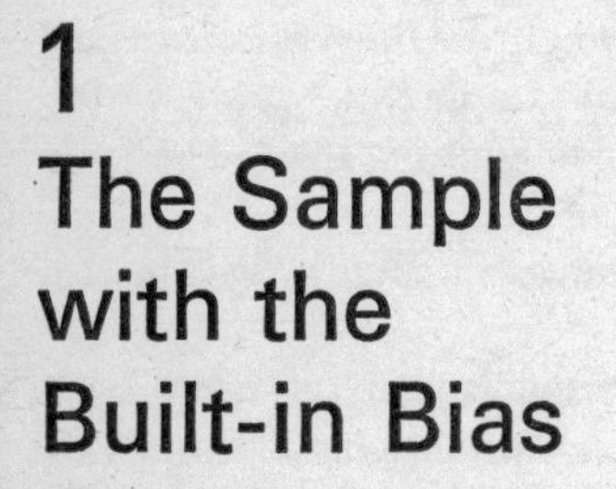

1
The Sample with the Built-in Bias

If you have a barrel of beans, some red and some white, there is only one way to find out precisely how many of each colour you have: Count 'em.

There is an easier way to discover *about* how many are red. Pull out a handful of beans and count just those, assuming that the proportion will be the same all through the barrel. If your sample is large enough and selected properly, it will represent the whole well enough for most purposes. If, however, it fails in either respect it may be far less accurate than an intelligent guess and have nothing to recommend it except a spurious air of scientific precision. It is sad truth that conclusions from such samples, biased by the method of selection, or too small, or both, lie behind much of what we read or think we know.

How a sample develops bias is most easily seen by looking

at an extreme example. Suppose you were to send to a group of your fellow-citizens a questionnaire that included this query: 'Do you like to answer questionnaires?' Add up the returns and you would very probably be able to announce that an overwhelming majority – which, for greater conviction, you would specify right down to the last decimal – of 'a typical cross-section of the population' asserts affection for the things. What has happened, of course, is that most of those whose answer would have been No have eliminated themselves from your sample by flinging your questionnaire into the nearest wastebasket. Even if the flingers constituted nine out of ten in your original sample you would be following a time-hallowed practice in ignoring them when you announced your findings.

Do such samples bias themselves in such a way in real life? You bet they do.

Newspapers and news magazines told us a while back that some four million American Catholics had become Protestants in the last ten years. Source was a poll conducted by the Reverend Daniel A. Poling, editor of the interdenominational *Christian Herald*. *Time* sums up the story:

> The *Herald* got its figures by polling a cross section of U.S. Protestant ministers, The 2,219 clergymen who replied to its questionnaire (out of 25,000 polled) reported that they had received a total of 51,361 former Roman Catholics into their churches within the past ten years. Projecting his sample, Poling got a nationwide estimate of 4,144,366 Catholic-to-Protestant converts in a decade. Writes Episcopalian Will Oursler: 'Even when allowances are made for error, the total national figure could hardly be less than two or three million and in all probability runs closer to five million.

Although it missed a bet in failing to point out the significance of the fact, *Time* deserves a small bow for letting us know that more than 90 per cent of the ministers polled did not reply. To destroy this survey completely, we have only to note the reasonable possibility that most of the 90 per cent threw away the questionnaire because they had no conversions to report.

Employing this assumption and using the same figure – 181,000 – as Dr Poling did for the total number of Protestant ministers with pastoral charges, we can make our own projection. Since he went to 25,000 out of 181,000 and found 51,361 conversions, asking everybody should have produced a conversion total of 370,000 or so.

Our crude methods have produced a very dubious figure, but it is at least as worthy of trust as the one that was published nationally – one that is eleven times as big as ours and therefore far more exciting.

As for Mr O's confident 'allowances . . . for error', well, if he has discovered a method to compensate for errors of unknown magnitude, the world of statistics will be grateful.

With this background, let us work over a news report – from some years back when it represented even more money than it does now – that 'the average Yale man, Class of '24, makes $25,111 a year'.

Well, good for him!

But wait a minute. What does this impressive figure mean? Is it, as it appears to be, evidence that if you send your boy to Yale, or, for all I know, Oxbridge, you won't have to work in your old age and neither will he?

Two things about the figure stand out at first suspicious glance. It is surprisingly precise. It is quite improbably salubrious.

There is small likelihood that the average income of any far-flung group is ever going to be known down to the dollar. It is not particularly probable that you know your own income for last year so precisely as that unless it was all derived from salary. And $25,000 incomes are not often all salary; people in that bracket are likely to have well-scattered investments.

Furthermore, this lovely average is undoubtedly calculated from the amounts the Yale men *said* they earned. Even if they had the honour system in New Haven in '24, we cannot be sure that it works so well after a quarter of a century that all these reports are honest ones. Some people when asked their incomes exaggerate out of vanity or op-

timism. Others minimize, especially, it is to be feared, on income-tax returns; and having done this may hesitate to contradict themselves on any other paper. Who knows what the revenuers may see? It is possible that these two tendencies, to boast and to understate, cancel each other out, but it is unlikely. One tendency may be far stronger than the other, and we do not know which one.

We have begun then to account for a figure that common sense tells us can hardly represent the truth. Now let us put our finger on the likely source of the biggest error, a source that can produce $25,111 as the 'average income' of some men whose actual average may well be nearer half that amount.

The report on the Yale men comes from a sample. We can be pretty sure of that because reason tells us that no one can get hold of all the living members of that class of '24. There are bound to be many whose addresses are unknown twenty-five years later.

And, of those whose addresses are known, many will not reply to a questionnaire, particularly a rather personal one. With some kinds of mail questionnaire, a five or ten per cent response is quite high. This one should have done better than that, but nothing like one hundred per cent.

So we find that the income figure is based on a sample composed of all class members whose addresses are known and who replied to the questionnaire. Is this a representative sample? That is, can this group be assumed to be equal in income to the unrepresented group, those who cannot be reached or who do not reply?

Who are the little lost sheep down in the Yale rolls as 'address unknown'? Are they the big-income earners – the Wall Street men, the corporation directors, the manufacturing and utility executives? No; the addresses of the rich will not be hard to come by. Many of the most prosperous members of the class can be found through *Who's Who in America* and other reference volumes even if they have neglected to keep in touch with the alumni office. It is a good guess that the lost names are those of the men who, twenty-five years or so after becoming Yale bachelors of arts, have not fulfilled any shining promise. They are clerks, mechanics, tramps, unemployed alcoholics, barely surviving writers and artists . . . people of whom it would take half a dozen or more to add up to an income of $25,111. These men do not so often register at class reunions, if only because they cannot afford the trip.

Who are those who chucked the questionnaire into the nearest wastebasket? We cannot be so sure about these, but it is at least a fair guess that many of them are just not making enough money to brag about. They are a little like the fellow who found a note clipped to his first pay cheque suggesting that he consider the amount of his salary confidential and not material for the interchange of office confidences. 'Don't worry,' he told the boss. 'I'm just as ashamed of it as you are.'

It becomes pretty clear that the sample has omitted two groups most likely to depress the average. The $25,111 figure is beginning to explain itself. If it is a true figure for anything it is one merely for that special group of the class of '24 whose addresses are known and who are willing to stand up and tell how much they earn. Even that requires an assumption that the gentlemen are telling the truth.

Such an assumption is not to be made lightly. Experience from one breed of sampling study, that called market research, suggests that it can hardly ever be made at all. A house-to-house survey purporting to study magazine readership was once made in which a key question was: What magazine does your household read? When the results were tabulated and analysed it appeared that a great many people loved *Harper's*, which if not highbrow is at least upper middlebrow, and not very many read *True Story*, which is very lowbrow indeed. Yet there were publishers' figures around at the time that showed very clearly that *True Story* had more millions of circulation than *Harper's* had hundreds of thousands. Perhaps we asked the wrong kind of people, the designers of the survey said to themselves. But no, the questions had been asked in all sorts of neighbourhoods all around the country. The only reasonable conclusion then was that a good many of the respondents, as people are called when they answer such questions, had not told the truth. About all the survey had uncovered was snobbery.

In the end it was found that if you wanted to know what certain people read it was no use asking them. You could learn a good deal more by going to their houses and saying you wanted to buy old magazines and what could be had?

Then all you had to do was count the *Yale Reviews* and the *Love Romances*. Even that dubious device, of course, does not tell you what people read, only what they have been exposed to.

Similarly, the next time you learn from your reading that the average man (you hear a good deal about him these days, most of it faintly improbable) brushes his teeth 1·02 times a day – a figure I have just made up, but it may be as good as anyone else's – ask yourself a question. How can anyone have found out such a thing? Is a woman who has read in countless advertisements that non-brushers are social offenders going to confess to a stranger that she does not brush her teeth regularly? The statistic may have meaning to one who wants to know only what people say about tooth-brushing but it does not tell a great deal about the frequency with which bristle is applied to incisor.

A river cannot, we are told, rise above its source. Well, it can seem to if there is a pumping station concealed somewhere about. It is equally true that the result of a sampling study is no better than the sample it is based on. By the time the data have been filtered through layers of statistical manipulation and reduced to a decimal-pointed average, the result begins to take on an aura of conviction that a closer look at the sampling would deny.

To be worth much, a report based on sampling must use a representative sample, which is one from which every source of bias has been removed. That is where our Yale figure shows its worthlessness. It is also where a great many of the things you can read in newspapers and magazines reveal their inherent lack of meaning.

A psychiatrist reported once that practically everybody is neurotic. Aside from the fact that such use destroys any meaning in the word 'neurotic', take a look at the man's sample. That is, whom has the psychiatrist been observing? It turns out that he has reached this edifying conclusion from studying his patients, who are a long, long way from being a sample of the population. If a man were normal, our psychiatrist would never meet him.

Give that kind of second look to the things you read, and you can avoid learning a whole lot of things that are not so.

It is worth keeping in mind also that the dependability of a sample can be destroyed just as easily by invisible sources of bias as by these visible ones. That is, even if you can't find a source of demonstrable bias, allow yourself some degree of scepticism about the result as long as there is a possibility of bias somewhere. There always is. The American presidential elections in 1948 and 1952 were enough to prove that, if there were any doubt.

For further evidence go back to 1936 and the *Literary Digest*'s famed fiasco. The ten million telephone and *Digest*

subscribers who assured the editors of the doomed magazine that it would be Landon 370, Roosevelt 161 came from the list that had accurately predicted the 1932 election. How could there be bias in a list already so tested? There was a bias, of course, as college theses and other post mortems found: People who could afford telephones and magazine subscriptions in 1936 were not a cross section of voters. Economically they were a special kind of people, a sample biased because it was loaded with what turned out to be Republican voters. The sample elected Landon, but the voters thought otherwise.

The basic sample is the kind called 'random'. It is selected by pure chance from the 'universe', a word by which the statistician means the whole of which the sample is a part. Every tenth name is pulled from a file of index cards. Fifty slips of paper are taken from a hatful. Every twentieth person met in Piccadilly is interviewed. (But remember that this last is not a sample of the population of the world, or of England, or of San Francisco, but only of the people in Piccadilly at the time. One interviewer for an opinion poll said that she got her people in a railroad station because 'all kinds of people can be found in a station'. It had to be pointed out to her that mothers of small children, for instance, might be under-represented there.)

The test of the random sample is this: Does every name or thing in the whole group have an equal chance to be in the sample?

The purely random sample is the only kind that can be examined with entire confidence by means of statistical theory, but there is one thing wrong with it. It is so difficult and expensive to obtain for many uses that sheer cost eliminates it. A more economical substitute, which is almost universally used in such fields as opinion polling and market research, is called stratified random sampling.

To get this stratified sample you divide your universe into several groups in proportion to their known prevalence. And right there your trouble can begin: Your information about their proportion may not be correct. You instruct your interviewers to see to it that they talk to so many Negroes and such-and-such a percentage of people in each of several income brackets, to a specified number of farmers, and so on. All the while the group must be divided equally between persons over forty and under forty years of age.

That sounds fine – but what happens? On the question of Negro or white the interviewer will judge correctly most of the time. On income he will make more mistakes. As to farmers – how do you classify a man who farms part time and works in the city too? Even the question of age can pose some problems which are most easily settled by choosing only respondents who obviously are well under or well over forty. In that case the sample will be biased by the virtual absence of the late-thirties and early-forties age groups. You can't win.

On top of all this, how do you get a random sample within the stratification? The obvious thing is to start with a list of everybody and go after names chosen from it at random; but that is too expensive. So you go into the streets – and bias your sample against stay-at-homes. You go from door to door by day – and miss most of the employed people. You switch to evening interviews – and neglect the movie-goers and night-clubbers.

The operation of a poll comes down in the end to a running battle against sources of bias, and this battle is conducted all the time by all the reputable polling organizations. What the reader of the reports must remember is that the battle is never won. No conclusion that 'sixty-seven per cent of the British people are against' something or other should be read without the lingering question. Sixty-seven per cent of which British people?

So with the late Dr Alfred C. Kinsey's 'male volume' and 'female volume'. Splendid ground-breakers though they have proved to be, they are cursed by sampling that is distressingly far from random. It is bad enough that the samples list heavily in such peculiar directions as college educated (seventy-five per cent of the women) and prison residence. A more serious weakness because harder to allow for is the probability that the samples lean sharply towards sexual exhibitionists; for folks who volunteer to tell all when the subject is sex may well differ sharply in their sexual histories from the more taciturn who have weeded themselves out of the samples by saying the hopeful interviewers nay.

That all this is more than speculation is confirmed by a study made by A. H. Maslow at Brooklyn College. Among girl students in his sample were many who later volunteered for kinseying, and Maslow found that these girls were generally the more sexually unconventional and sexually sophisticated ones.

The problem when reading Kinsey, or any of the more recent studies of sexual behaviour for that matter, is how to study it without learning too much that is not necessarily so. The danger is acute with any research based on sampling, and it is likely to become even more so when you take your big book or major research report in the form of a popular summary.

For one thing, there are at least three levels of sampling involved in work like Kinsey's. As already noted, the samples of the population (first level) are far from random and so may not be particularly representative of any population. It is equally important to remember that any questionnaire is only a sample (another level) of the possible questions; and that the answer the gentleman or lady gives is no more than a sample (third level) of his or her attitudes and experiences on each question.

It may be true with the Kinsey kind of work, as it has been found to be elsewhere, that the kind of people who make up an interviewing staff can shade the results in an interesting fashion. Several skirmishes back, sometime during World War Two, the National Opinion Research

Center sent out two staffs of interviewers to ask three questions of five hundred Negroes in a Southern city of the United States. White interviewers made up one staff, black the other.

One question was, 'Would Negroes be treated better or worse here if the Japanese conquered the U.S.A.?' Negro interviewers reported that nine per cent of those they asked said 'better'. White interviewers found only two per cent of such responses. And while Negro interviewers found only twenty-five per cent who thought Negroes would be treated worse, white interviewers turned up forty-five per cent.

When 'Nazis' was substituted for 'Japanese' in the question, the results were similar.

The third question probed attitudes that might be based on feelings revealed by the first two. 'Do you think it is more important to concentrate on beating the Axis, or to make democracy work better here at home?' 'Beat the Axis' was the reply of thirty-nine per cent, according to the Negro interviewers; of sixty-two per cent, according to the white.

Here is bias introduced by unknown factors. It seems likely that the most effective factor was a tendency that must always be allowed for in reading polls results, a desire to give a pleasing answer. Would it be any wonder if, when answering a question with connotations of disloyalty in wartime, a Southern Negro would tell a white man what sounded good rather than what he actually believed? It is also possible that the different groups of interviewers chose different kinds of people to talk to.

In any case the results are obviously so biased as to be worthless. You can judge for yourself how many other poll-based conclusions are just as biased, just as worthless – but with no check available to show them up.

You have pretty fair evidence to go on if you suspect that polls in general are biased in one specific direction, the direction of the *Literary Digest* error. This bias is towards the person with more money, more education, more information and alertness, better appearance, more conventional behaviour, and more settled habits than the average of the population he is chosen to represent.

You can easily see what produces this. Let us say that you are an interviewer assigned to a street corner, with one interview to get. You spot two men who seem to fit the category you must complete: over forty, Negro, urban. One is in clean overalls, decently patched, neat. The other is dirty and he looks surly. With a job to get done, you approach the more likely-looking fellow, and your colleagues all over the country are making similar decisions.

Some of the strongest feeling against public-opinion polls is found in liberal or left-wing circles, where it is rather commonly believed that polls are generally rigged. Behind this view is the fact that poll results so often fail to square with the opinions and desires of those whose thinking is not in the conservative direction. Polls, they point out, seem to elect Republicans even when voters shortly thereafter do otherwise.

Actually, as we have seen, it is not necessary that a poll be rigged – that is, that the results be deliberately twisted in order to create a false impression. The tendency of the sample to be biased in this consistent direction can rig it automatically.

2 The Well-Chosen Average

You, I trust, are not a snob, and I certainly am not an estate agent. But let's say that you are and that I am and that you are looking for property to buy along a road I know well. Having sized you up, I take pains to tell you that the average income in this neighbourhood is some £10,000 a year. Maybe that clinches your interest in living here; anyway, you buy and that handsome figure sticks in your mind. More than likely, since we have agreed that for the purposes of the moment you are a bit of a snob, you toss it in casually when telling your friends about where you live.

A year or so later we meet again. As a member of some rate-payers' committee I am circulating a petition to keep the rates down or assessments down or bus fare down. My plea is that we cannot afford the increase: After all, the average income in this neighbourhood is only £2,000 a year.

Perhaps you go along with me and my committee in this – you're not only a snob, you're stingy too – but you can't help being surprised to hear about that measly £2,000. Am I lying now, or was I lying last year?

You can't pin it on me either time. That is the essential beauty of doing your lying with statistics. Both those figures are legitimate averages, legally arrived at. Both represent the same data, the same people, the same incomes. All the same it is obvious that at least one of them must be so misleading as to rival an out-and-out lie.

My trick was to use a different kind of average each time, the word 'average' having a very loose meaning. It is a trick commonly used, sometimes in innocence but often in guilt, by fellows wishing to influence public opinion or sell advertising space. When you are told that something is an average you still don't know very much about it unless you can find out which of the common kinds of average it is – mean, median, or mode.

The £10,000 figure I used when I wanted a big one is a mean, the arithmetic average of the incomes of all the families in the neighbourhood. You get it by adding up all the incomes and dividing by the number there are. The smaller figure is a median, and so it tells you that half the families in question have more than £2,000 a year and half have less. I might also have used the mode, which is the most frequently met-with figure in a series. If in this neighbourhood there are more families with incomes of £3,000 a year than with any other amount, £3,000 a year is the modal income.

In this case, as usually is true with income figures, an unqualified 'average' is virtually meaningless. One factor that adds to the confusion is that with some kinds of information all the averages fall so close together that, for casual purposes, it may not be vital to distinguish among them.

If you read that the average height of the men of some

primitive tribe is only five feet, you get a fairly good idea of the stature of these people. You don't have to ask whether that average is a mean, median, or mode; it would come out about the same. (Of course, if you are in the business of manufacturing overalls for Africans you would want more information than can be found in any average. This has to do with ranges and deviations, and we'll tackle that one in the next chapter.)

The different averages come out close together when you deal with data, such as those having to do with many human characteristics, that have the grace to fall close to what is called the normal distribution. If you draw a curve to represent it you get something shaped like a bell, and mean, median, and mode fall at the same point.

Consequently one kind of average is as good as another for describing the heights of men, but for decribing their pocketbooks it is not. If you should list the annual incomes of all the families in a given city you might find that they ranged from not much to perhaps £20,000 or so, and you might find a few very large ones. More than nine-five per cent of the incomes would be under £5,000, putting them way over towards the left-hand side of the curve. Instead of being symmetrical, like a bell, it would be skewed. Its shape would be a little like that of a child's slide, the ladder rising sharply to a peak, the working part sloping gradually down. The mean would be quite a distance from the median. You can see what this would do to the validity of any comparison made between the 'average' (mean) of one year and the 'average' (median) of another.

In the neighbourhood where I sold you some property the two averages are particularly far apart because the distribution is markedly skewed. It happens that most of your neighbours are small farmers or wage earners employed in a near-by village or elderly retired people on pensions. But three of the inhabitants are millionaire week-enders and these three boost the total income, and therefore the arithmetical average, enormously. They boost it to a figure that practically everybody in the neighbourhood has a good deal less than. You have in reality the case that sounds like a joke or a figure of speech: Nearly everybody is below average.

That's why when you read an announcement by a corporation executive or a business proprietor that the average pay of the people who work in his establishment is so much, the figure may mean something and it may not. If the average is a median, you can learn something significant from it: Half the employees make more than that; half make less. But if it is a mean (and believe me it may be that if its nature is unspecified) you may be getting nothing more revealing than the average of one £25,000 income – the proprietor's – and the salaries of a crew of underpaid workers. 'Average annual pay of £3,800' may conceal both the £1,400 salaries and the owner's profits taken in the form of a whopping salary.

How neatly this can be worked into a whipsaw device, in which the worse the story, the better it looks, is illustrated in some company statements. Let's try our hand at one in a small way.

You are one of the three partners who own a small manufacturing business. It is now the end of a very good year. You have paid out £99,000 to the ninety employees who do the work of making and shipping the chairs or whatever it is that you manufacture. You and your partners have paid yourselves £5,500 each in salaries. You find there are profits for the year of £21,000 to be divided equally among you. How are you going to describe this? To make it easy to understand, you put it in the form of averages. Since all the employees are doing about the same kind of work for

25,000

7,600

5,500

3,472 Arithmetical average

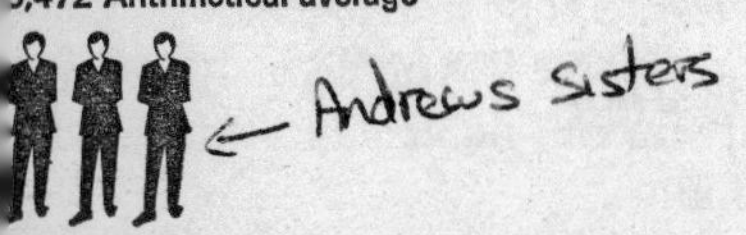

3,500

2,500

2,100 Median (the one in the middle. 12 above him, 12 below)

1,400

Mode (occurs most frequently)

similar pay it won't make much difference whether you use a mean or a median. This is what you come out with:

Average wage of employees	£1,100
Average salary and profit of owners	12,500

That looks terrible, doesn't it? Let's try it another way. Take £15,000 of the profits and distribute it among the three partners as bonuses. And this time when you average up the wages, include yourself and your partners. And be sure to use a mean.

Average wage or salary	£1,403
Average profit of owners	2,000

Ah. That looks better. Not as good as you could make it look, but good enough. Less than six per cent of the money available for wages and profits has gone into profits, and you can go further and show that too if you like. Anyway, you've got figures now that you can publish, post on a bulletin board, or use in bargaining.

This is pretty crude because the example is simplified, but it is nothing to what has been done in the name of accounting. Given a complex corporation with hierarchies of employees ranging all the way from beginning typist to president with a several-hundred-thousand-dollar bonus, all sorts of things can be covered up in this manner.

So when you see an average-pay figure, first ask: Average of what? Who's included? The United States Steel Corporation once said that its employees' average weekly earnings went up 107 per cent in less than a decade. So they did – but some of the punch goes out of the magnificent increase when you note the earlier figure includes a much larger number of partially employed people. If you work half-time one year and full-time the next, your earnings will double, but that doesn't indicate anything at all about your wage rate.

You may have read in the paper that the income of the average American family was $6,940 in some specified year. You should not try to make too much out of that figure unless you also know what 'family' has been used to mean, as well as what kind of average this is. (And who says so and how he knows and how accurate the figure is.)

The figure you saw may have come from the Bureau of the Census. If you have the Bureau's full report you'll have no trouble finding right there the rest of the information you need: that this average is a median; that 'family' signifies 'two or more persons related to each other and living together'. You will also learn, if you turn back to the tables, that the figure is based on a sample of such size that there are nineteen chances out of twenty that the estimate is correct within a margin of, say, $71 plus or minus.

That probability and that margin add up to a pretty good estimate. The Census people have both skill enough and money enough to bring their sampling studies down to a fair degree of precision. Presumably they have no particular axes to grind. Not all the figures you see are born under such happy circumstances, nor are all of them accompanied by any information at all to show how precise or imprecise they may be. We'll work that one over in the next chapter.

Meanwhile you may want to try your scepticism on some items from 'A Letter from the Publisher' in *Time* magazine. Of new subscribers it said, 'Their median age is 34 years and their average family income is $7,270 a year.' An earlier survey of 'old TIMERs' had found that their 'median age was 41 years . . . Average income was $9,535 . . .' The natural question is why, when median is given for ages both times, the kind of average for incomes is carefully unspecified.

Could it be that the mean was used instead because it is bigger, thus seeming to dangle a richer readership before advertisers?

You might also try a game of what-kind-of-average-are-you on the alleged prosperity of the 1924 Yales reported at the beginning of Chapter 1.

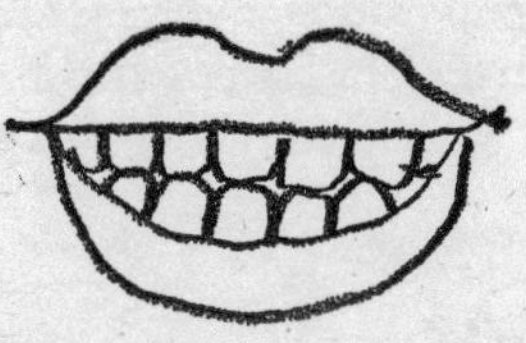

3 The Little Figures That Are Not There

What you should do when told the results of a survey, a statistician once advised, is ask, 'How many juries did you poll before you found this one?'

As noted previously, well-biased samples can be employed to produce almost any result anyone may wish. So can properly random ones, if they are small enough and you try enough of them.

Users report 23 per cent fewer cavities with Doakes' toothpaste, the big type says. You could do with twenty-three per cent fewer aches so you read on. These results, you find, come from a reassuringly 'independent' laboratory, and the account is certified by a chartered accountant. What more do you want?

Yet if you are not outstandingly gullible or optimistic, you will recall from experience that one toothpaste is seldom much better than any other. Then how can the Doakes people report such results? Can they get away with telling lies, and in such big type at that? No, and they don't have to. There are easier ways and more effective ones.

The principal joker in this one is the inadequate sample – statistically inadequate, that is; for Doakes' purpose it is just right. That test group of users, you discover by reading the small type, consisted of just a dozen persons. (You have to hand it to Doakes, at that, for giving you a sporting chance.

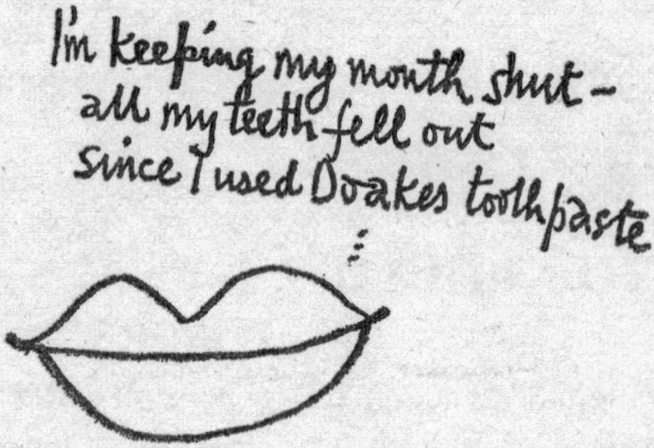

Some advertisers would omit this information and leave even the statistically sophisticated only a guess as to what species of chicanery was afoot. His sample of a dozen isn't so bad either, as these things go. Something called Dr Cornish's Tooth Powder came onto the market a few years ago with a claim to have shown 'considerable success in correction of . . . dental caries'. The idea was that the powder contained urea, which laboratory work was supposed to have demonstrated to be valuable for the purpose. The pointlessness of this was that the experimental work had been purely preliminary and had been done on precisely six cases.)

But let's get back to how easy it is for Doakes to get a headline without a falsehood in it and everything certified at that. Let any small group of persons keep count of cavities for six months, then switch to Doakes'. One of three things is bound to happen: distinctly more cavities, distinctly fewer, or about the same number. If the first or last of these

possibilities occurs, Doakes & Company files the figures (well out of sight somewhere) and tries again. Sooner or later, by the operation of chance, a test group is going to show a big improvement worthy of a headline and perhaps a whole advertising campaign. This will happen whether they adopt Doakes' or baking soda or just keep on using their same old dentifrice.

The importance of using a small group is this: With a large group any difference produced by chance is likely to be a small one and unworthy of big type. A two-per-cent-improvement claim is not going to sell much toothpaste.

How results that are not indicative of anything can be produced by pure chance – given a small enough number of cases – is something you can test for yourself at small cost. Just start tossing a penny. How often will it come up heads? Half the time, of course. Everyone knows that.

Well, let's check that and see ... I have just tried ten tosses and got heads eight times, which proves that pennies come up heads eighty per cent of the time. Well, by toothpaste statistics they do. Now try it yourself. You may get a fifty-fifty result, but probably you won't; your result, like mine, stands a good chance of being quite a way away from fifty-fifty. But if your patience holds out for a thousand tosses you are almost (though not quite) certain to come out with a result very close to half heads – a result, that is, which represents the real probability. Only when there is a substantial number of trials involved is the law of averages a useful description or prediction.

How many is enough? That's a tricky one too. It depends among other things on how large and how varied a population you are studying by sampling. And sometimes the number in the sample is not what it appears to be.

A remarkable instance of this came out in connection with a test of a polio vaccine some years ago. It appeared to be an impressively large-scale experiment as medical ones go: 450 children were vaccinated in a community and 680 were left unvaccinated, as controls. Shortly thereafter the community was visited by an epidemic. Not one of the vaccinated children contracted a recognizable case of polio.

Neither did any of the controls. What the experimenters had overlooked or not understood in setting up their project was the low incidence of paralytic polio. At the usual rate, only two cases would have been expected in a group of this size, and so the test was doomed from the start to have no meaning. Something like fifteen to twenty-five times this many children would have been needed to obtain an answer signifying anything.

Many a great, if fleeting, medical discovery has been launched similarly. 'Make haste', as one physician* put it, 'to use a new remedy before it is too late.'

The guilt does not always lie with the medical profession alone. Public pressure and hasty journalism often launch a treatment that is unproved, particularly when the demand is great and the statistical background hazy. So it was with the cold vaccines that were popular some years back and the

* These words have been attributed to both Sir William Osler and Edward Livingston Trudeau. Choose one. Since they were both physicians and pretty sharp on our subject, it is quite possible that they both said it, give or take a word or two.

antihistamines more recently. A good deal of the popularity of these unsuccessful 'cures' sprang from the unreliable nature of the ailment and from a defect of logic. Given time, a cold will cure itself.

How can you avoid being fooled by inconclusive results? Must every man be his own statistician and study the raw data for himself? It is not that bad; there is a test of significance that is easy to understand. It is simply a way of reporting how likely it is that a test figure represents a real result rather than something produced by chance. This is the little figure that is not there – on the assumption that you, the lay reader, wouldn't understand it. Or that, where there's an axe to grind, you would.

If the source of your information gives you also the degree of significance, you'll have a better idea of where you stand. This degree of significance is most simply expressed as a probability, as when the Bureau of the Census tells you that there are nineteen chances out of twenty that their figures have a specified degree of precision. For most purposes nothing poorer than this five per cent level of significance is good enough. For some the demanded level is one per cent, which means that there are ninety-nine chances out of a hundred that an apparent difference, or whatnot, is real. Anything this likely is sometimes described as 'practically certain'.

There's another kind of little figure that is not there, one whose absence can be just as damaging. It is the one that tells the range of things or their deviation from the average that is given. Often an average – whether mean or median, specified or unspecified – is such an oversimplification that it is worse than useless. Knowing nothing about a subject is frequently healthier than knowing what is not so, and a little learning may be a dangerous thing.

Altogether too much housing, for instance, has been

planned to fit the statistically average family of 3·6 persons. Translated into reality this means three or four persons, which, in turns, means two bedrooms. And this size family, 'average' though it is, actually makes up a minority of all families. 'We build average houses for average families,' say the builders – and neglect the majority that are larger or smaller. Some areas, in consequence of this, have been over-

built with two-bedroom houses, underbuilt in respect to smaller and larger units. So here is a statistic whose misleading incompleteness has had costly consequences. Of it a large public-health group has said: 'When we look beyond the arithmetical average to the actual range which it misrepresents, we find that the three-person and four-person families make up only 45 per cent of the total. Thirty-five per cent are one-person and two-person; 20 per cent have more than four persons.'

Common sense has somehow failed in the face of the convincingly precise and authoritative 3·6. It has somehow outweighed what everybody knows from observation: that many families are small and quite a few are large.

In somewhat the same fashion those little figures that are missing from what are called 'Gesell's norms' have produced pain in papas and mamas. Let a parent read, as many have done in such places as Sunday papers, that 'a child' learns to sit erect at the age of so many months and he thinks at once of his own child. Let his child fail to sit by the specified age and the parent must conclude that his offspring is 'retarded' or 'subnormal' or something equally invidious. Since half the children are bound to fail to sit by the time mentioned, a good many parents are made unhappy. Of course, speaking mathematically, this unhappiness is balanced by the joy of the other fifty per cent of parents in discovering that their children are 'advanced'. But harm can come of the efforts of the unhappy parents to force their children to conform to the norms and thus be backward no longer.

All this does not reflect on Dr Arnold Gesell or his methods. The fault is in the filtering-down process from the researcher through the sensational or ill-informed writer to the reader who fails to miss the figures that have disappeared in the process. A good deal of the misunderstanding can be avoided if to the 'norm' or average is added an indication of the range. Parents seeing that their youngsters fall within the normal range will quit worrying about small and meaningless differences. Hardly anybody is exactly normal in any way, just as one hundred tossed pennies will rarely come up exactly fifty heads and fifty tails.

Confusing 'normal' with 'desirable' makes it all the worse. Dr Gesell simply stated some observed facts; it was the parents who, in reading the books and articles, concluded that a child who walks late by a day or a month must be inferior.

A good deal of the sillier criticism of Dr Alfred Kinsey's well-known (if hardly well-read) report came from taking normal to be equivalent to good, right, desirable. Dr Kinsey was accused of corrupting youth by giving them ideas and particularly by calling all sorts of popular but unapproved sexual practices normal. But he simply said that he had found these activities to be usual, which is what normal means, and he did not stamp them with any seal of approval. Whether they were naughty or not did not come within what Dr Kinsey considered to be his province. So he ran up against something that has plagued many another observer: It is dangerous to mention any subject having high emotional content without hastily saying whether you are for or agin it.

The deceptive thing about the little figure that is not there is that its absence so often goes unnoticed. That, of course, is the secret of its success. Critics of journalism as practised today have deplored the paucity of good old-fashioned leg work and spoken harshly of 'armchair correspondents', who live by uncritically re-writing government handouts. For a sample of unenterprising journalism take this item from a list of 'new industrial developments' in the news magazine *Fortnight*: 'a new cold temper bath which triples the hardness of steel, from Westinghouse'.

Now that sounds like quite a development . . . until you try to put your finger on what it means. And then it becomes as elusive as a ball of quicksilver. Does the new bath make just any kind of steel three times as hard as it was before treatment? Or does it produce a steel three times as hard as any previous steel? Or what does it do? It appears that the reporter has passed along some words without inquiring what they mean, and you are expected to read them just as uncritically for the happy illusion they give you of having learned something. It is all too reminiscent of an old definition of the lecture method of classroom instruction: a process by which the contents of the textbook of the instructor are transferred to the notebook of the student without passing through the heads of either party.

A few minutes ago, while looking up something about Dr Kinsey in *Time*, I came upon another of those statements that collapse under a second look. It appeared in an advertisement by a group of electric companies in 1948. 'Today, electric power is available to more than three-quarters of U.S. farms. . . .' That sounds pretty good. Those power companies are really on the job. Of course, if you wanted to be ornery you could paraphrase it into 'Almost one-quarter of U.S. farms do not have electric power available today.' The real gimmick, however, is in that word 'available', and by using it the companies have been able to say just about anything they please. Obviously this does not mean that all those farmers actually have power, or the advertisement surely would have said so. They merely have it 'available' – and that, for all I know, could mean that the power lines go past their farms or merely within ten or a hundred miles of them.

Let me quote a title from an article published in a popular magazine: 'You Can Tell *Now* HOW TALL YOUR CHILD WILL GROW.' With the article is conspicuously displayed a pair of charts, one for boys and one for girls, showing what percentage of his ultimate height a child reaches at each year of age. 'To determine your child's height at maturity,' says a caption, 'check present measurement against chart.'

The funny thing about this is that the article itself – if you read on – tells you what the fatal weakness in the chart is. Not all children grow in the same way. Some start slowly

Worldwide availability of 'How to Lie with Statistics'

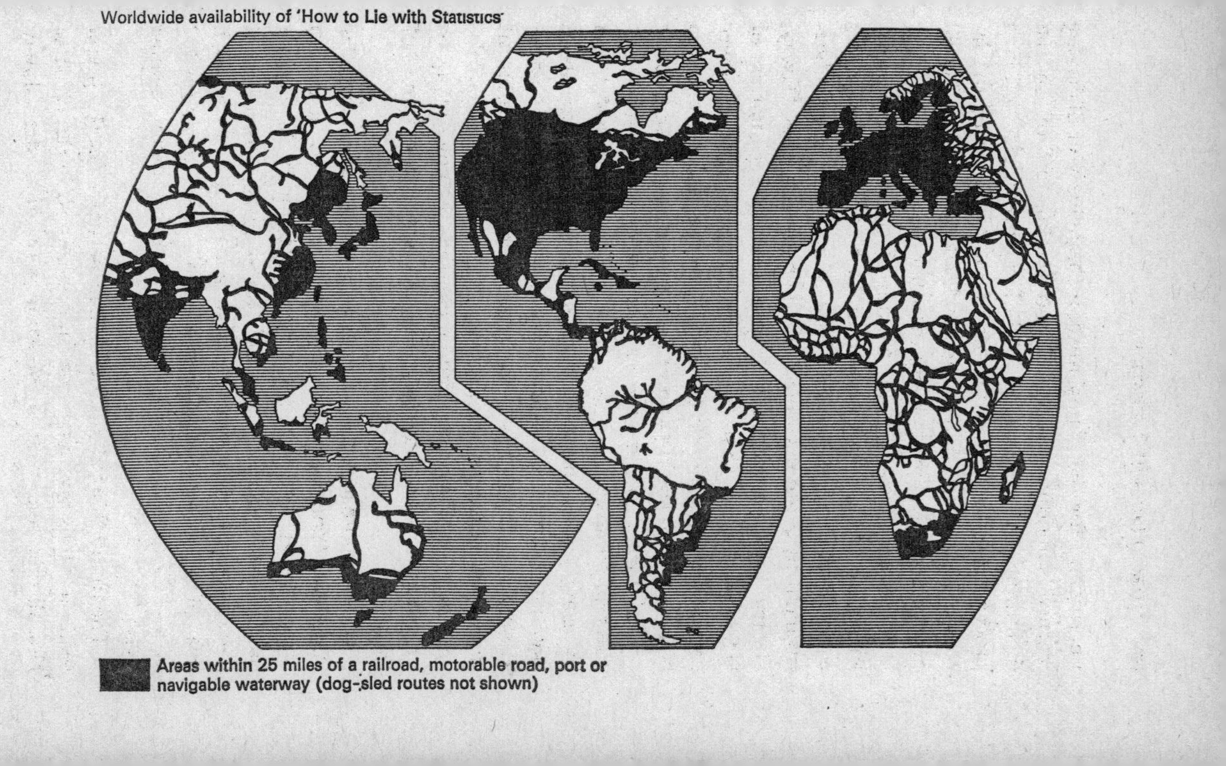

and then speed up; others shoot up quickly for a while, then level off slowly; for still others growth is a relatively steady process. The chart, as you might guess, is based on averages taken from a large number of measurements. For the total, or average, heights of a hundred youngsters taken at random it is no doubt accurate enough, but a parent is interested in only one height at a time, a purpose for which such a chart is virtually worthless. If you wish to know how tall your child is going to be, you can probably make a better

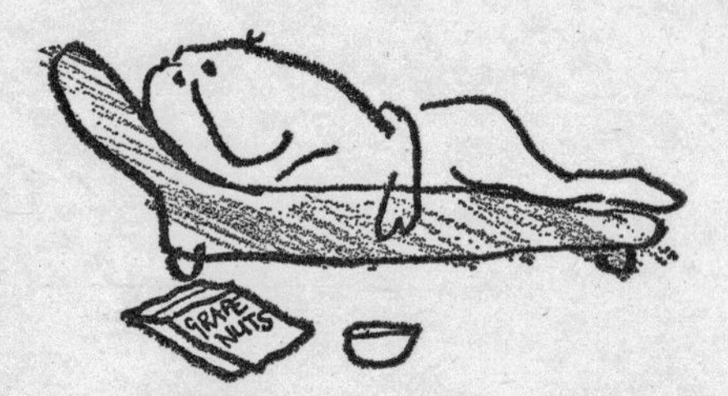

guess by taking a look at his parents and grandparents. That method isn't scientific and precise like the chart, but it is at least as accurate.

I am amused to note that, taking my height as recorded when I enrolled in high-school military training at fourteen and ended up in the rear rank of the smallest squad, I should eventually have grown to a bare five feet eight. I am five feet eleven. A three-inch error in human height comes down to a poor grade of guess.

Before me are wrappers from two boxes of Grape-Nuts Flakes. They are slightly different editions, as indicated by their testimonials: one cites Two-Gun Pete and the other says, 'If you want to be like Hoppy . . . you've got to eat like Hoppy!' Both offer charts to show ('Scientists *proved* it's true!') that these flakes 'start giving you energy in 2 minutes!' In one case the chart hidden in these forests of exclamation points has numbers up the side; in the other case the numbers have been omitted. This is just as well,

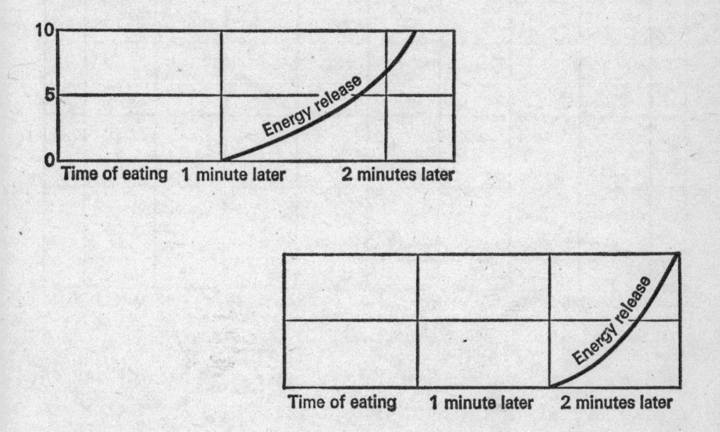

since there is no hint of what the numbers mean. Both show a steeply climbing red line ('energy release'), but one has it starting one minute after eating Grape-Nuts Flakes, the other two minutes later. One line climbs about twice as fast as the other, too, suggesting that even the draughtsman didn't think these graphs meant anything.

Such foolishness could be found only on material meant for the eye of a juvenile or his morning-weary parent, of course. No one would insult a big businessman's intelligence with such statistical tripe . . . or would he? Let me tell you about a graph used to advertise an advertising agency (I hope this isn't getting confusing) in the rather special columns of *Fortune* magazine. The line on this graph showed the impressive upward trend of the agency's business year by year. There were no numbers. With equal honesty this chart could have represented a tremendous

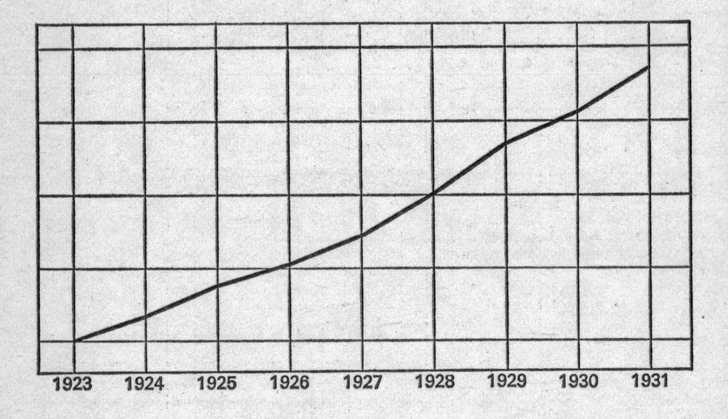

growth, with business doubling or increasing by millions of dollars a year, or the snail-like progress of a static concern adding only a dollar or two to its annual billings. It made a striking picture, though.

Place little faith in an average or a graph or a trend when those important figures are missing. Otherwise you are as blind as a man choosing a camp site from a report of mean temperature alone. You might take 61 degrees as a comfortable annual mean, giving you a choice in California between such areas as the inland desert and San Nicolas Island off the south coast. But you can freeze or roast if you ignore the range. For San Nicolas it is 47 to 87 degrees but for the desert it is 15 to 104.

Oklahoma City can claim a similar average temperature for the last sixty years: 60·2 degrees. But as you can see from the chart below, that cool and comfortable figure conceals a range of 130 degrees.

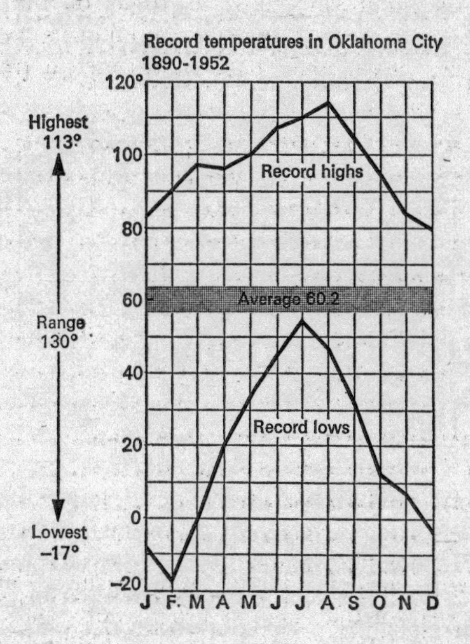

4
Much Ado about Practically Nothing

Sir Josiah Stamp has described an occasion when Lord Randolph was examining a report of revenue. His private secretary was looking over his shoulder. Randolph remarked that it was gratifying to find customs revenue up 34 per cent over a corresponding period of the previous year.

The secretary corrected him, pointing out that it was only ·34 per cent.

'What difference does that make?' Lord Randolph asked.

When it had been explained that one figure was a hundred times the other, Randolph said, 'I have often seen those damned little dots before, but I never knew until now what they meant.'

Not dots but other damned little differences crop up to plague comparisons of test scores. To see how this is, we will begin – if you don't mind – by endowing you with two children. Peter and Linda (we might as well give them modish names while we're about it) have been given intelligence tests, as a great many children are in the course of their schooling. Now the mental test of any variety is one of the prime voodoo fetishes of our time, so you may have to argue a little to find out the results of the tests; this is information so esoteric that it is often held to be safe only in the hands of psychologists and educators, and they may be right at that. Anyway, you learn somehow that Peter's IQ is 98 and Linda's is 101. You know, of course, that the IQ is based on 100 as average or 'normal'.

Aha. Linda is your brighter child. She is, furthermore, above average. Peter is below average, but let's not dwell on *that.*

Any such conclusions as these are sheer nonsense.

Just to clear the air, let's note first of all that whatever an intelligence test measures it is not quite the same thing as we usually mean by intelligence. It neglects such important things as leadership and creative imagination. It takes no account of social judgement or musical or artistic or other aptitudes, to say nothing of such personality matters as diligence and emotional balance. On top of that, the tests most often given in schools are the quick-and-cheap group kind that depend a good deal upon reading facility; bright or not, the poor reader hasn't a chance.

Let's say that we have recognized all that and agree to regard the IQ simply as a measure of some vaguely defined capacity to handle canned abstractions. And Peter and Linda have been given what is generally regarded as the best of the tests, the Revised Stanford-Binet, which is administered individually and doesn't call for any particular reading ability.

Now what an IQ test purports to be is a sampling of the

intellect. Like any other product of the sampling method, the IQ is a figure with a statistical error, which expresses the precision or reliability of that figure.

Asking these test questions is something like what you might do in estimating the quality of the maize in a field by going about and pulling off an ear here and an ear there at random. By the time you had stripped down and looked at a hundred ears, say, you would have gained a pretty good idea of what the whole field was like. Your information would be exact enough for use in comparing this field with another field – provided the two fields were not very similar. If they were, you might have to look at many more ears, rating them all the while by some precise standard of quality.

How accurately your sample can be taken to represent the whole field is a measure that can be represented in figures: the probable error and the standard error.

Suppose that you had the task of measuring the size of a good many fields by pacing off the fence lines. The first thing you might do is check the accuracy of your measuring system by pacing off what you took to be a hundred yards, doing this a number of times. You might find that on the average you were off by three yards. That is, you came within three yards of hitting the exact one hundred in half your trials, and in the other half of them you missed by more than three yards.

Your probable error then would be three yards in one hundred, or three per cent. From then on, each fence line that measured one hundred yards by your pacing might be recorded as 100 ± 3 yards.

(Most statisticians now prefer to use another, but comparable, measurement called the standard error. It takes in about two-thirds of the cases instead of exactly half and is considerably handier in a mathematical way. For our purposes we can stick to the probable error, which is the one still used in connection with the Stanford-Binet.)

As with our hypothetical pacing, the probable error of the Stanford-Binet IQ has been found to be three per cent. This has nothing to do with how good the test is basically, only with how consistently it measures whatever it measures. So Peter's indicated IQ might be more fully expressed as 98 ± 3 and Linda's as 101 ± 3.

This says that there is no more than an even chance that Peter's IQ falls anywhere between 95 and 101; it is just as likely that it is above or below that figure. Similarly Linda's has no better than a fifty-fifty probability of being within the range of 98 to 104. From this you can quickly see that there is one chance in four that Peter's IQ is really above 101 and a similar chance that Linda's is below 98. Then Peter is not inferior but superior, and by a margin of anywhere from three points up.

What this comes down to is that the only way to think about IQs and many other sampling results is in ranges. 'Normal' is not 100, but the range of 90 to 110, say, and there would be some point in comparing a child in this range with a child in a lower or higher range. But comparisons between figures with small differences are meaningless. You must always keep that plus-or-minus in mind, even (or especially) when it is not stated.

Ignoring these errors, which are implicit in all sampling studies, has led to some remarkably silly behaviour. There are magazine editors to whom readership surveys are gospel, mainly because they do not understand them. With forty per cent male readership reported for one article and only thirty-five per cent for another, they demand more articles like the first.

The difference between thirty-five and forty per cent readership can be of importance to a magazine, but a survey difference may not be a real one. Costs often hold readership samples down to a few hundred persons, particularly after those who do not read the magazine at all have been eliminated. For a magazine that appeals primarily to women the number of men in the sample may be very small. By the time these have been divided among those who say they 'read all', 'read most', 'read some', or 'didn't read' the article in question, the thirty-five per cent conclusion may be based on only a handful. The probable error hidden behind the impressively presented figure may be so large that the editor who relies on it is grasping at a thin straw.

Sometimes the big ado is made about a difference that is mathematically real and demonstrable but so tiny as to have no importance. This is in defiance of the fine old saying that a difference is a difference only if it makes a difference. A case in point is the hullabaloo over practically nothing that was raised so effectively, and so profitably, by the Old Gold cigarette people.

It started innocently with the editor of the *Reader's Digest*, who smokes cigarettes but takes a dim view of them all the same. His magazine went to work and had a battery of laboratory folk analyse the smoke from several brands of cigarettes. The magazine published the results, giving the nicotine and whatnot content of the smoke by brands. The conclusion stated by the magazine and borne out in its detailed figures was that all the brands were virtually identical and that it didn't make any difference which one you smoked.

Now you might think this was a blow to cigarette manufacturers and to the fellows who think up the new copy angles in the advertising agencies. It would seem to explode all advertising claims about soothing throats and kindness to T-zones.

But somebody spotted something. In the lists of almost

identical amounts of poisons, one cigarette had to be at the bottom, and the one was Old Gold. Out went the telegrams, and big advertisements appeared in newspapers at once in the biggest type at hand. The headlines and the copy simply said that of all cigarettes tested by this great national magazine Old Gold had the least of these undesirable things in its smoke. Excluded were all figures and any hint that the difference was negligible.

In the end, the Old Gold people were ordered to 'cease and desist' from such misleading advertising. That didn't make any difference; the good had been milked from the idea long before.

5 The Gee-Whiz Graph

There is terror in numbers. Humpty Dumpty's confidence in telling Alice that he was master of the words he used would not be extended by many people to numbers. Perhaps we suffer from a trauma induced by early experiences with maths.

Whatever the cause, it creates a real problem for the writer who yearns to be read, the advertising man who expects his copy to sell goods, the publisher who wants his books or magazines to be popular. When numbers in tabular form are taboo and words will not do the work well, as is often the case, there is one answer left: Draw a picture.

About the simplest kind of statistical picture, or graph, is the line variety. It is very useful for showing trends, something practically everybody is interested in showing or knowing about or spotting or deploring or forecasting.

We'll let our graph show how national income increased ten per cent in a year.

Begin with paper ruled into squares. Name the months along the bottom. Indicate billions of dollars up the side. Plot your points and draw your line, and your graph will look like this:

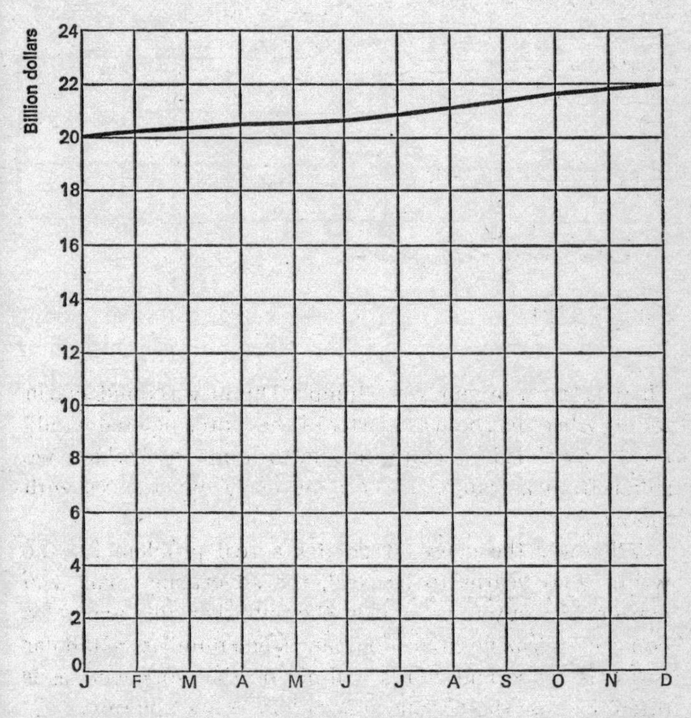

Now that's clear enough. It shows what happened during the year and it shows it month by month. He who runs may see and understand, because the whole graph is in proportion and there is a zero line at the bottom for comparison. Your ten per cent *looks* like ten per cent – an upward trend that is substantial but perhaps not overwhelming.

That is very well if all you want to do is convey information. But suppose you wish to win an argument, shock a reader, move him into action, sell him something. For that, this chart lacks schmaltz. Chop off the bottom.

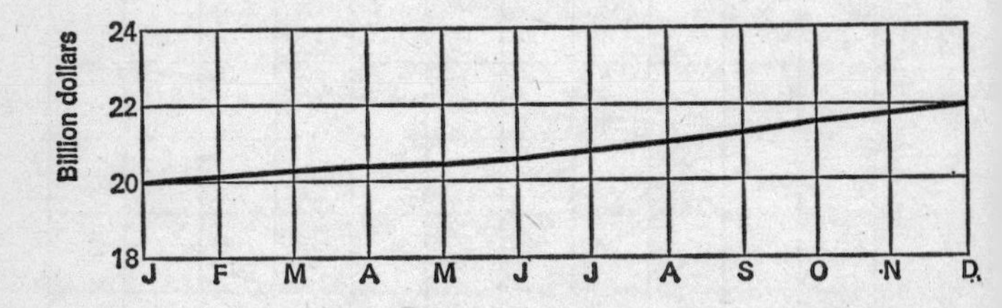

Now that's more like it. (You've saved paper too, something to point out if any carping fellow objects to your misleading graphics.) The figures are the same and so is the curve. It is the same graph. Nothing has been falsified – except the impression that it gives. But what the hasty reader sees now is a national-income line that has climbed half-way up the paper in twelve months, all because most of the chart isn't there any more. Like the missing parts of speech in sentences that you met in grammar classes, it is 'understood'. Of course, the eye doesn't 'understand' what isn't there, and a small rise has become, visually, a big one.

Now that you have practised to deceive, why stop with truncating? You have a further trick available that's worth a dozen of that. It will make your modest rise of ten per cent look livelier than one hundred per cent is entitled to look. Simply change the proportion between the ordinate and the abscissa. There's no rule against it, and it does give your graph a prettier shape. All you have to do is let each mark up the side stand for only one-tenth as many dollars as before.

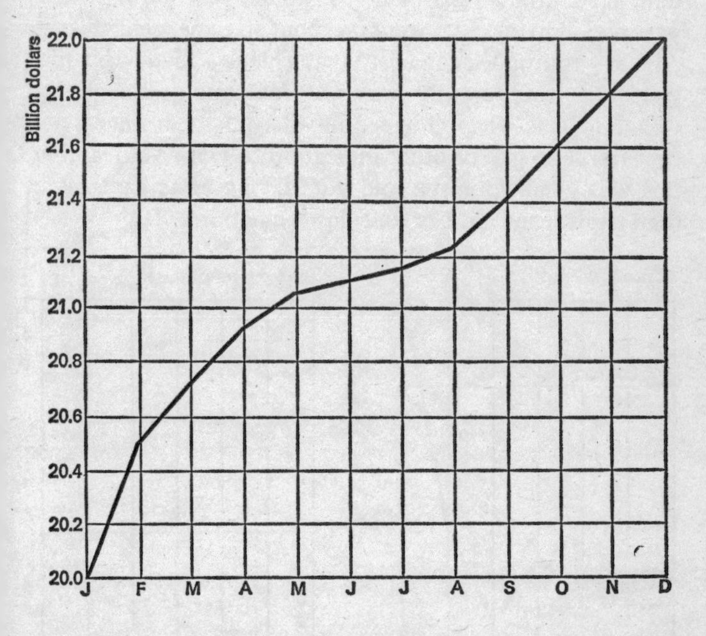

That *is* impressive, isn't it? Anyone looking at it can just feel prosperity throbbing in the arteries of the country. It is a subtler equivalent of editing 'National income rose ten per cent' into '... climbed a whopping ten per cent'. It is vastly more effective, however, because it contains no adjectives or adverbs to spoil the illusion of objectivity. There's nothing anyone can pin on you.

And you're in good, or at least respectable, company. A news magazine has used this method to show the stock market hitting a new high, the graph being so truncated as to make the climb look far more dizzying than it was. A

Columbia Gas System advertisement once reproduced a chart 'from our new Annual Report'. If you read the little numbers and analysed them you found that during a ten-year period living costs went up about sixty per cent and the cost of gas dropped four per cent. This is a favourable picture, but it apparently was not favourable enough for Columbia Gas. They chopped off their chart at ninety per cent (with no gap or other indication to warn you) so that this was what your eye told you: Living costs have more than tripled, and gas has gone down one-third!

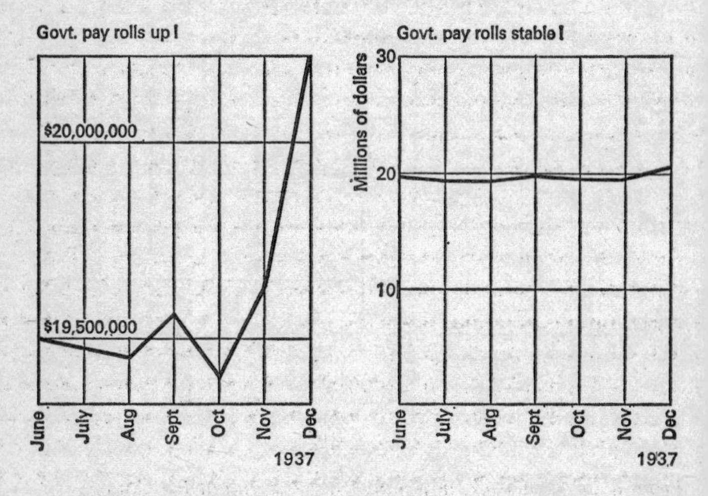

Steel companies have used similarly misleading graphic methods in attempts to line up public opinion against wage increases. Yet the method is far from new, and its impropriety was shown up long ago – not just in technical publications for statisticians either. An editorial writer in *Dun's Review* back in 1938 reproduced a chart from an advertisement advocating advertising in Washington, D.C., the argument being nicely expressed in the headline over the chart: GOVERNMENT PAY ROLLS UP! The line in the graph went along with the exclamation point even though the figures behind it did not. What they showed was an increase from about $19,500,000 to $20,200,000. But the red line shot from near the bottom of the graph clear to the top, making an increase of under four per cent look like more than 400. The magazine gave its own graphic version of the same figures alongside – an honest red line that rose just four per cent, under this caption: GOVERNMENT PAY ROLLS STABLE.

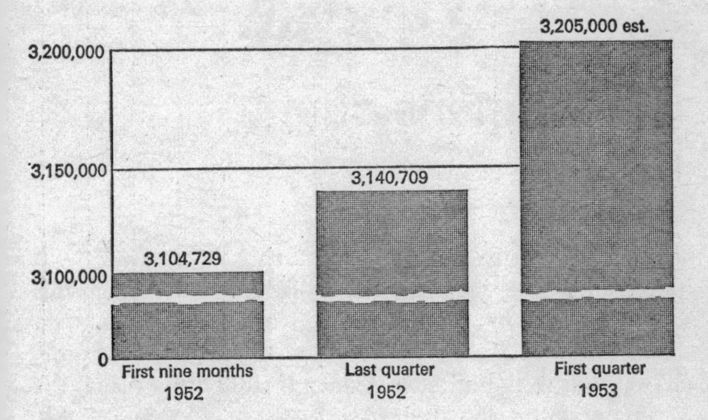

From a 24 April 1953, newspaper advertisement for Collier's.

6
The One-Dimensional Picture

A generation or so ago we were hearing a good deal about the little people, meaning practically all of us. When this began to sound too condescending, we became the common man. Pretty soon that was forgotten too, which was probably just as well. But the little man is still with us. He is the character on the chart.

A chart on which a little man represents a million men, a moneybag or stack of coins a thousand pounds sterling or a million dollars, an outline of a steer your beef supply for next year, is a pictorial graph. It is a useful device. It has what I am afraid is known as eye-appeal. And it is capable of becoming a fluent, devious, and successful liar.

The daddy of the pictorial chart, or pictograph, is the ordinary bar chart, a simple and popular method of representing quantities when two or more are to be compared.

A bar chart is capable of deceit too. Look with suspicion on any version in which the bars change their widths as well as their lengths while representing a single factor or in which they picture three-dimensional objects the volumes of which are not easy to compare. A truncated bar chart has, and deserves, exactly the same reputation as the truncated line graph we have been talking about. The habitat of the bar chart is the geography book, the corporation statement, and the news magazine. This is true also of its eye-appealing offspring.

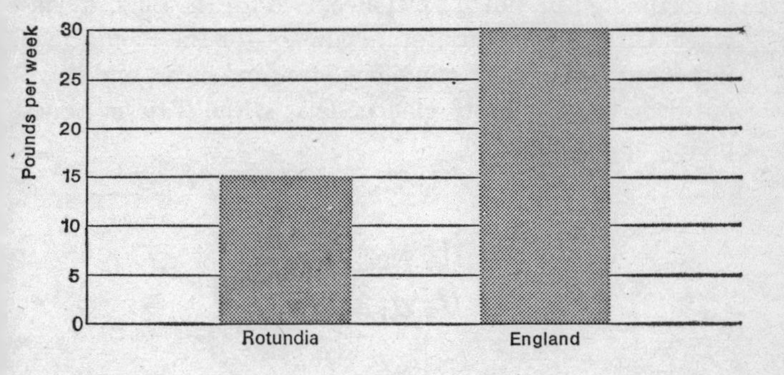

Perhaps I wish to show a comparison of two figures – the average weekly wage of one kind of working man or another in England and Rotundia, let's say. The sums might be £30 and £15. I wish to catch your eye with this, so I am not satisfied merely to print the numbers. I make a bar chart. (By the way, if that £30 figure doesn't square with the huge sum you laid out when your porch needed a new railing last summer, remember that your man may not have done as well every week as he did while working for you. And anyway I didn't say what kind of average I have in mind or how I arrived at it, so it isn't going to get you anywhere to quibble. You see how easy it is to hide behind the most disreputable statistic if you don't include any other information with it? You probably guessed I just made this one up for purposes of illustration, but I'll bet you wouldn't have if I'd used £29.35 instead.)

There it is, with pounds-per-week indicated up the left side. It is a clear and honest picture. Twice as much money is twice as big on the chart and looks it.

The chart lacks that eye-appeal though, doesn't it? I can easily supply that by using something that looks more like money than a bar does: moneybags. One moneybag for the unfortunate Rotundian's pittance, two for the Englishman's wage. Or three for the Rotundian, six for the Englishman. Either way, the chart remains honest and clear, and it will not deceive your hasty glance. That is the way an honest pictograph is made.

That would satisfy me if all I wanted was to communicate information. But I want more. I want to say that the English working man is vastly better off than the Rotundian, and the more I can dramatize the difference between fifteen and thirty the better it will be for my argument. To tell the truth (which, of course, is what I am planning not to do), I want you to infer something, to come away with an exaggerated impression, but I don't want to be caught at my tricks. There is a way, and it is one that is being used every day to fool you.

I simply draw a moneybag to represent the Rotundian's £15, and then I draw another one twice as tall to represent the Englishman's £30. That's in proportion, isn't it?

Now *that* gives the impression I'm after. The Englishman's wage now dwarfs the foreigner's.

The catch, of course, is this. Because the second bag is twice as high as the first, it is also twice as wide. It occupies not twice but four times as much area on the page. The numbers still say two to one, but the visual impression, which is the dominating one most of the time, says the ratio is four to one. Or worse. Since these are pictures of objects having in reality three dimensions, the second must also be twice as thick as the first. As your geometry book put it, the volumes of similar solids vary as the cube of any like dimension. Two times two times two is eight. If one moneybag holds £15, the other, having eight times the volume, must hold not £30 but £120.

And that indeed is the impression my ingenious little chart gives. While saying 'twice', I have left the lasting impression of an overwhelming eight-to-one ratio.

You'll have trouble pinning any criminal intent on me, too. I am only doing what a great many other people do. A leading news magazine has done it repeatedly, with moneybags just like those in our example.

In America, the Iron and Steel Institute has done it, with a pair of blast furnaces. The idea was to show how the industry's steelmaking capacity had boomed between one decade and the next and so indicate that the industry was doing such a job on its own that any governmental interference was uncalled for. However shaky the thesis, it has more

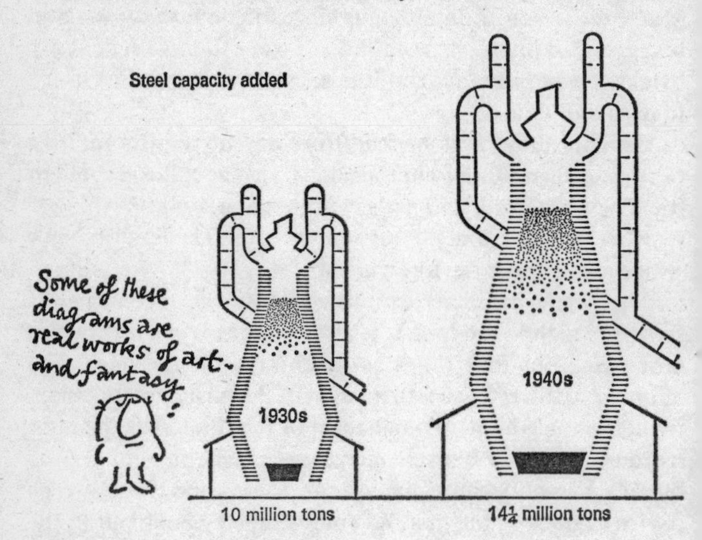

Adapted by courtesy of Steelways.

merit than there is in the way it was presented. The blast furnace representing the ten-million-ton capacity added in one decade was drawn just over two-thirds as tall as the furnace representing the fourteen and a quarter million tons added in the next. The eye saw two furnaces, one of them close to three times as big as the other. To say 'almost one and one-half' and to be heard as 'three' – that's what the one dimensional picture can accomplish.

This piece of art work by the steel people had some other

points of interest. Somehow the second furnace had fattened out horizontally beyond the proportion of its neighbour, and a black bar, suggesting molten iron, had become two and one-half times as long as in the earlier decade. Here was a 50 per cent increase given, then drawn as 150 per cent to give a visual impression of – unless my slide rule and I are getting out of their depth – over 1,500 per cent. Arithmetic becomes fantasy.

(It is almost too unkind to mention that the same glossy four-colour page offers a fair-to-prime specimen of the truncated line graph. A curve exaggerates the per-capita growth of steelmaking capacity by getting along with the lower half of its graph missing. This saves paper and doubles the rate of climb.)

Some of this may be no more than sloppy draughtsmanship. But it is rather like being short-changed: When all the mistakes are in the cashier's favour, you can't help wondering.

Newsweek once showed how 'Old Folks Grow Older' by means of a chart on which appeared two male figures, one representing the 68·2-year life expectancy of today, the other the 34-year life expectancy of 1879–89. It was the same old story: One figure was twice as tall as the other and so would have had eight times the bulk or weight. This picture sensationalized facts in order to make a better story. I would call it a form of yellow journalism. The same issue of the magazine contained a truncated, or gee-whiz, line graph.

There is still another kind of danger in varying the size of objects in a chart. It seems that in 1860 there were something over eight million milk cows in the United States and about a century later there were more than twenty-five million. Showing this increase by drawing two cows, one three times the height of the other, will exaggerate the

The crescive cow

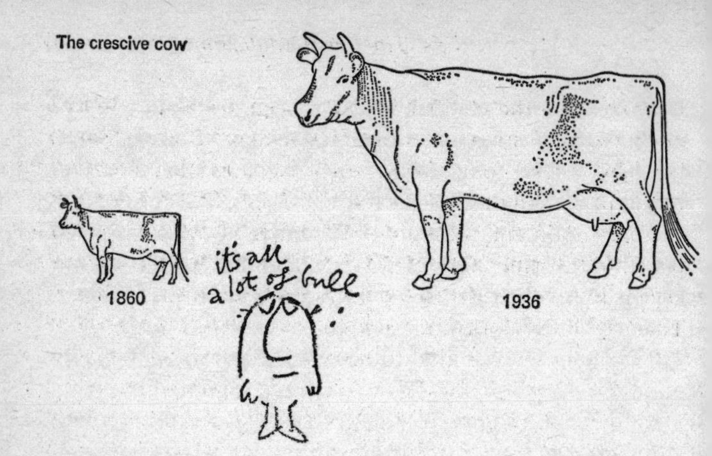

The diminishing rhinoceros

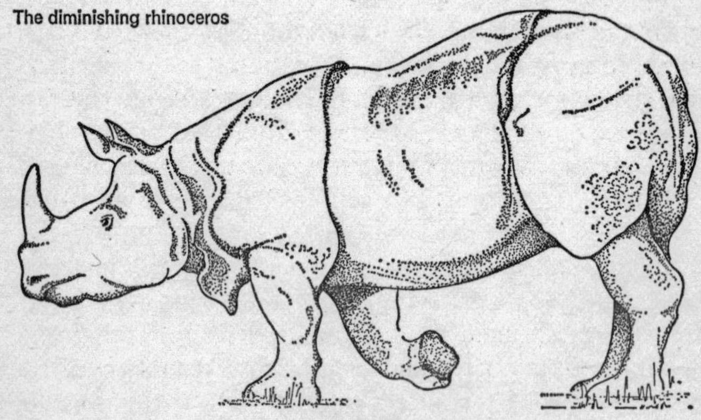

impression in the manner we have been discussing. But the effect on the hasty scanner of the page may be even stranger: He may easily come away with the idea that cows are bigger now than they used to be.

Apply the same deceptive technique to what has happened to the rhinoceros population and this is what you get. Ogden Nash once rhymed rhinosterous with preposterous. That's the word for the method too.

7 The Semi-attached Figure

'When you are a bit older,' a judge in India once told an eager young British civil servant, 'you will not quote Indian statistics with that assurance. The government are very keen on amassing statistics – they collect them, add them, raise them to the *n*th power, take the cube root and prepare wonderful diagrams. But what you must never forget is that every one of those figures comes in the first instance from the *chowty dar* [village watchman], who just puts down what he damn pleases.'

If you can't prove what you want to prove, demonstrate something else and pretend that they are the same thing. In the daze that follows the collision of statistics with the human mind, hardly anybody will notice the difference. The semi-attached figure is a device guaranteed to stand you in good stead. It always has.

You can't prove that your nostrum cures colds, but you can publish (in large type) a sworn laboratory report that half an ounce of the stuff killed 31,108 germs in a test tube in eleven seconds. While you are about it, make sure that the laboratory is reputable or has an impressive name. Reproduce the report in full. Photograph a doctor-type model in white clothes and put his picture alongside.

80% of known germs use people

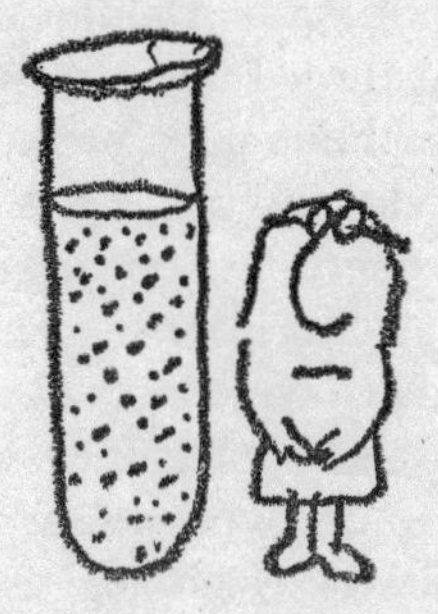

But don't mention the several gimmicks in your story. It is not up to you – is it? – to point out that an antiseptic that works well in a test tube may not perform in the human throat, especially after it has been diluted according to instructions to keep it from burning throat tissue. Don't confuse the issue by telling what kind of germ you killed. Who knows what germ causes colds, particularly since it probably isn't a germ at all?

In fact, there is no known connection between assorted

germs in a test tube and the whatever-it-is that produces colds, but people aren't going to reason that sharply, especially while sniffling.

Maybe that one is too obvious, and people are beginning to catch on, although it would not appear so from the advertising pages. Anyway, here is a trickier version.

Let us say that during a period in which race prejudice is growing you are employed to 'prove' otherwise. It is not a difficult assignment. Set up a poll or, better yet, have the polling done for you by an organization of good reputation. Ask that usual cross section of the population if they think black people have as good a chance as white people to get jobs. Repeat your polling at intervals so that you will have a trend to report.

Princeton's Office of Public Opinion Research tested this question once. What turned up is interesting evidence that things, especially in opinion polls, are not always what they seem. Each person who was asked the question about jobs was also asked some questions designed to discover if he was strongly prejudiced against blacks. It turned out that people most strongly prejudiced were most likely to answer Yes to the question about job opportunities. (It worked out that about two-thirds of those who were sympathetic towards Negroes did not think the Negro had as good a chance at a job as a white person did, and about two-thirds of those showing prejudice said that blacks were getting as good breaks as whites.) It was pretty evident that from this poll you would learn very little about employment conditions for Negroes, although you might learn some interesting things about a man's racial attitudes.

You can see, then, that if prejudice is mounting during your polling period you will get an increasing number of answers to the effect that Negroes have as good a chance at jobs as whites. So you announce your results: Your poll shows that blacks are getting a fairer shake all the time.

You have achieved something remarkable by careful use of a semi-attached figure. The worse things get, the better your poll makes them look.

Or take this one: '27 per cent of a large sample of eminent physicians smoke Throaties – more than any other brand.' The figure itself may be phoney, of course, in any of several

ways, but that really doesn't make any difference. The only answer to a figure so irrelevant is 'So what?' With all proper respect towards the medical profession, do doctors know any more about tobacco brands than you do? Do they have any inside information that permits them to choose the least harmful among cigarettes? Of course they don't, and your doctor would be the first to say so. Yet that '27 per cent' somehow manages to sound as if it meant something.

Now slip back one percentage point and consider the case of the juice extractor. It was widely advertised as a device that 'extracts 26 per cent more juice' as 'proved by laboratory test' and 'vouched for by Good Housekeeping Institute'.

That sounds right good. If you can buy a juicer that is twenty-six per cent more effective, why buy any other kind? Well now, without going into the fact that 'laboratory tests' (especially 'independent laboratory tests') have proved some of the darndest things, just what does that figure mean? Twenty-six per cent more than what? When it was finally pinned down it was found to mean only that this juicer got out that much more juice than an old-fashioned hand reamer could. It had absolutely nothing to do with the data you would want before purchasing; this juicer might be the poorest on the market. Besides being suspiciously precise, that twenty-six per cent figure is totally irrelevant.

Advertisers aren't the only people who will fool you with numbers if you let them. An article on driving safety, published by *This Week* magazine undoubtedly with your best interests at heart, told you what might happen to you if you went 'hurtling down the highway at 70 miles an hour, careering from side to side.' You would have, the article said, four times as good a chance of staying alive if the time were seven in the morning than if it were seven at night. The evidence: 'Four times more fatalities occur on the highways at 7 p.m. than at 7 a.m.' Now that is approximately true, but the conclusion doesn't follow. More people are killed in the evening than in the morning mainly because more people are on the highways then to be killed. You, a single driver, may be in greater danger in the evening, but there is nothing in the figures to prove it either way.

By the same kind of nonsense that the article writer used you can show that clear weather is more dangerous than foggy weather. More accidents occur in clear weather, because there is more clear weather than foggy weather. All the same, fog may be much more dangerous to drive in.

You can use accident statistics to scare yourself to death in connection with any kind of transportation ... if you fail to note how poorly attached the figures are.

More people were killed by aeroplanes last year than in 1910. Therefore modern planes are more dangerous? Nonsense. There are hundreds of times more people flying now, that's all.

It was reported that the number of deaths chargeable to railroads in one year was 4,712. That sounds like a good argument for staying off trains, perhaps for sticking to your automobile instead. But when you investigate to find what the figure is all about, you learn it means something quite different. Nearly half those victims were people whose automobiles collided with trains at crossings. The greater part of the rest were riding the rods. Only 132 out of the 4,712 were passengers on trains. And even that figure is worth little for purposes of comparison unless it is attached to information on total passenger miles.

If you are worried about your chances of being killed on a coast-to-coast trip, you won't get much relevant information by asking whether trains, planes, or cars killed the greatest number of people last year. Get the rate, by inquiring into the number of fatalities for each million passenger miles. That will come closest to telling you where your greatest risk lies.

There are many other forms of counting up something and then reporting it as something else. The general method is to pick two things that sound the same but are not. As personnel manager for a company that is scrapping with a union you 'make a survey' of employees to find out how many have a complaint against the union. Unless the union is a band of angels with an archangel at their head you can ask and record with perfect honesty and come out with proof that the greater part of the men do have some complaint or other. You issue your information as a report that 'a vast majority – 78 per cent – are opposed to the union.' What you have done is to add up a bunch of undifferentiated complaints and tiny gripes and then call them something else that sounds like the same thing. You haven't proved a thing, but it rather sounds as if you have, doesn't it?

It is fair enough, though, in a way. The union can just as readily 'prove' that practically all the workers object to the way the plant is being run.

If you'd like to go on a hunt for semi-attached figures, you might try running through corporation financial statements. Watch for profits that might look too big and so are concealed under another name. The United Automobile Workers' magazine *Ammunition* describes the device this way:

> The statement says, last year the company made $35 million in profits. Just one and a half cents out of every sales dollar. You feel sorry for the company. A bulb burns out in the latrine. To replace it, the company has to spend 30 cents. Just like that, there is the profit on 20 sales dollars. Makes a man want to go easy on the paper towels.
>
> But, of course, the truth is, what the company reports as profits is only a half or a third of the profits. The part that isn't reported is hidden in depreciation, and special depreciation, and in reserves for contingencies.

Equally gay fun is to be had with percentages. For a nine-month period General Motors was able to report a relatively modest profit (after taxes) of 12·6 per cent on sales. But for that same period GM's profit on its investment came to 44·8 per cent, which sounds a good deal worse – or better, depending on what kind of argument you are trying to win.

Similarly, a reader of *Harper's* magazine came to the defence of the A & P stores in that magazine's letters column by pointing to low net earnings of only 1·1 per cent of sales. He asked, 'Would any American citizen fear public condemnation as a profiteer ... for realizing a little over $10 for every $1,000 invested during a year?'

Offhand this 1·1 per cent sounds almost distressingly small. Compare it with the six per cent or more interest that most of us are familiar with from home mortgages and bank loans and such. Wouldn't the A & P be better off if it went out of the grocery business and put its capital into the bank and lived off interest?

The catch is that annual return on investment is not the same kettle of fish as earnings on total sales. As another reader replied in a later issue of *Harper's*, 'If I purchase an article every morning for 99 cents and sell it each afternoon for one dollar, I will make only 1 per cent on total sales, but 365 per cent on invested money during the year.'

There are often many ways of expressing any figure. You can, for instance, express exactly the same fact by calling it a one per cent return on sales, a fifteen per cent return on investment, a ten-million-dollar profit, an increase in profits of forty per cent (compared with 1965–9 average), or a decrease of sixty per cent from last year. The method is to choose the one that sounds best for the purpose at hand and trust that few who read it will recognize how imperfectly it reflects the situation.

Not all semi-attached figures are products of intentional deception. Many statistics, including medical ones that are pretty important to everybody, are distorted by inconsistent reporting at the source. There are startlingly contradictory figures on such delicate matters as abortions, illegitimate births, and syphilis. If you should look up the latest available figures on influenza and pneumonia in the United States, you might come to the strange conclusion that these ailments are practically confined to three southern states, which account for about eighty per cent of the reported cases. What actually explains this percentage is the fact that these three states required reporting of the ailments after other states had stopped doing so.

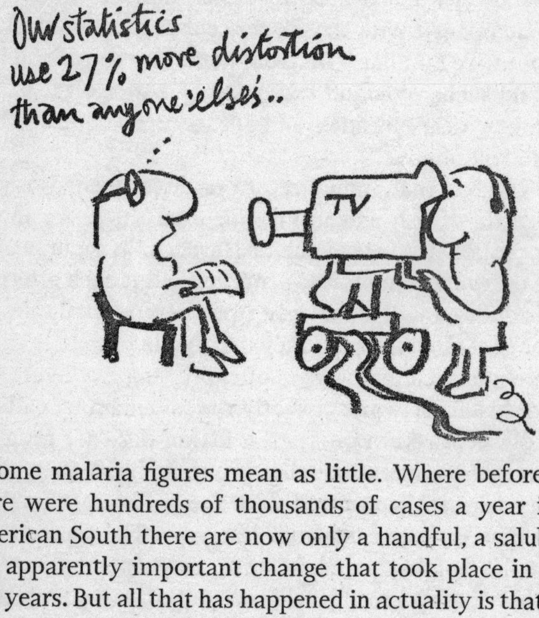

Some malaria figures mean as little. Where before 1940 there were hundreds of thousands of cases a year in the American South there are now only a handful, a salubrious and apparently important change that took place in just a few years. But all that has happened in actuality is that cases are now recorded only when proved to be malaria, where formerly the word was used in much of the South as a colloquialism for a cold or chill.

The death rate in the Navy during the Spanish–American War was nine per thousand. For civilians in New York City during the same period it was sixteen per thousand. Navy recruiters later used these figures to show that it was safer to be in the U.S. Navy than out of it. Assume these figures to be accurate, as they probably are. Stop for a moment and see if you can spot what makes them, or at least the conclusion the recruiting people drew from them, virtually meaningless.

The groups are not comparable. The Navy is made up mainly of young men in known good health. A civilian population includes infants, the old, and the ill, all of whom have a higher death rate wherever they are. These figures do not at all prove that men meeting Navy standards will live longer in the Navy than out. They do not prove the contrary either.

Shortly before polio vaccines came along we were hit by the discouraging news that the previous year had been the worst in history for polio. This conclusion was based on what might seem all the evidence anyone could ask for: There were far more cases reported in that year than ever before.

But when experts went back of these figures they found a few things that were more encouraging. One was that there were so many more children at the most susceptible ages than ever before that cases were bound to be at a record number if the rate remained level. Another was that a general consciousness of polio was leading to more frequent diagnoses and recording of mild cases. Finally, there was an increased financial incentive, there being more polio insurance and more aid available from charitable organizations. All this threw considerable doubt on the notion that polio had reached a new high, and the total number of deaths confirmed the doubt.

It is an interesting fact that the death rate or number of deaths often is a better measure of the incidence of an ailment than direct incidence figures – simply because the quality of reporting and record-keeping is so much higher on fatalities. In this instance, the obviously semi-attached figure is better than the one that on the face of it seems fully attached.

In America the semi-attached figure enjoys a big boom every fourth year. This indicates not that the figure is cyclical in nature, but only that campaign time has arrived. A campaign statement issued by the Republican party in October of 1948 is built entirely on figures that appear to be attached to each other but are not:

> When Dewey was elected Governor in 1942, the minimum teacher's salary in some districts was as low as $900 a year. Today the school teachers in New York State enjoy the highest salaries in the world. Upon Governor Dewey's recommendation, based on the findings of a Committee he appointed, the Legislature in 1947 appropriated $32,000,000 out of a state surplus to provide an immediate increase in the salaries of school teachers. As a result the minimum salaries of teachers in New York City range from $2,500 to $5,325.

It is entirely possible that Mr Dewey has proved himself the teacher's friend, but these figures don't show it. It is the old before-and-after trick, with a number of unmentioned factors introduced and made to appear what they are not. Here you have a 'before' of $900 and an 'after' of $2,500 to $5,325, which sounds like an improvement indeed. But the small figure is the lowest salary in any rural district of the state, and the big one is the range in New York City alone. There may have been an improvement under Governor Dewey, and there may not.

This statement illustrates a statistical form of the before-and-after photograph that is a familiar stunt in magazines and advertising. A living-room is photographed twice to show you what a vast improvement a coat of paint can make. But between the two exposures new furniture has been added, and sometimes the 'before' picture is a tiny one in poorly lighted black-and-white and the 'after' version is a big photograph in full colour. Or a pair of pictures shows you what happened when a girl began to use a hair rinse. By golly, she does look better afterwards at that. But most of the change, you note on careful inspection, has been wrought by persuading her to smile and throwing a back light on her hair. More credit belongs to the photographer than to the rinse.

8
Post Hoc Rides Again

You can make an estimate – one that is better than chance would produce – of how many children have been born into a Dutch or Danish family by counting the storks' nests on the roof of their house.

In statistical terminology it would be said that a positive correlation has been found to exist between these two things.

What sounds like proof of an ancient myth is actually something far more valuable. It is an easily remembered reminder of a useful truth: an association between two factors is not proof that one has caused the other.

In the instance of the storks and the babies, it is not too hard to find a third factor that may be responsible for the other two. Big houses attract big, and potentially big, families; and big houses have more chimney pots on which storks may nest.

But flaws in assumptions of causality are not always so easy to spot, especially when the relationship seems to make a lot of sense or when it pleases a popular prejudice.

Somebody once went to a good deal of trouble to find out if cigarette smokers make lower college grades than non-smokers. It turned out that they did. This pleased a good many people and they have been making much of it ever since. The road to good grades, it would appear, lies in giving up smoking; and, to carry the conclusion one reasonable step further, smoking makes dull minds.

This particular study was, I believe, properly done: sample big enough and honestly and carefully chosen, correlation having a high significance, and so on.

The fallacy is an ancient one which, however, has a powerful tendency to crop up in statistical material, where it is disguised by a welter of impressive figures. It is the one that says that if B follows A, then A has caused B. An unwarranted assumption is being made that since smoking and low grades go together, smoking causes low grades. Couldn't it just as well be the other way around? Perhaps low marks drive students not to drink but to tobacco. When it comes right down to it, this conclusion is about as likely as the other and just as well supported by the evidence. But it is not nearly so satisfactory to propagandists.

It seems a good deal more probable, however, that neither of these things has produced the other, but both are a product of some third factor. Can it be that the sociable sort of fellow who takes his books less than seriously is also likely to smoke more? Or is there a clue in the fact that somebody once established a correlation between extroversion and low grades – a closer relationship apparently than the one between grades and intelligence? Maybe extroverts smoke more than introverts. The point is that when there are many reasonable explanations you are hardly entitled to pick one that suits your taste and insist on it. But many people do.

To avoid falling for the *post hoc* fallacy and thus wind up believing many things that are not so, you need to put any statement of relationship through a sharp inspection. The correlation, that convincingly precise figure that seems to prove that something is because of something, can actually be any of several types.

One is the correlation produced by chance. You may be able to get together a set of figures to prove some unlikely thing in this way, but if you try again, your next set may not prove it at all. As with the manufacturer of the tooth paste that appeared to reduce decay, you simply throw away the results you don't want and publish widely those you do. Given a small sample, you are likely to find some substantial correlation between any pair of characteristics or events that you can think of.

A common kind of co-variation is one in which the relationship is real but it is not possible to be sure which of the variables is the cause and which the effect. In some of these instances cause and effect may change places from time to time or indeed both may be cause and effect at the same time. A correlation between income and ownership of stocks might be of that kind. The more money you make, the more stock you buy, and the more stock you buy, the more income you get; it is not accurate to say simply that one has produced the other.

Perhaps the trickiest of them all is the very common instance in which neither of the variables has any effect at all on the other, yet there is a real correlation. A good deal of dirty work has been done with this one. The poor grades among cigarette smokers is in this category, as are all too many medical statistics that are quoted without the qualification that although the relationship has been shown to be real, the cause-and-effect nature of it is only a matter of speculation. As an instance of the nonsense or spurious correlation that is a real statistical fact, someone has gleefully pointed to this: There is a close relationship between the salaries of Presbyterian ministers in Massachusetts and the price of rum in Havana.

Which is the cause and which the effect? In other words, are the ministers benefiting from the rum trade or supporting it? All right. That's so farfetched that it is ridiculous at a glance. But watch out for other applications of *post hoc* logic that differ from this one only in being more subtle. In the case of the ministers and the rum it is easy to see that both figures are growing because of the influence of a third factor: the historic and world-wide rise in the price level of practically everything.

And take the figures that show the suicide rate to be at its maximum in June. Do suicides produce June brides – or do June weddings precipitate suicides of the jilted? A somewhat more convincing (though equally unproved) explanation is that the fellow who licks his depression all through the winter with the thought that things will look rosier in the spring gives up when June comes and he still feels terrible.

Another thing to watch out for is a conclusion in which a correlation has been inferred to continue beyond the data with which it has been demonstrated. It is easy to show that the more it rains in an area, the taller the corn grows or even the greater the crop. Rain, it seems, is a blessing. But a season of very heavy rainfall may damage or even ruin the crop. The positive correlation holds up to a point and then quickly becomes a negative one. Above so-many inches, the more it rains the less corn you get.

A correlation of course shows a tendency which is not often the ideal relationship described as one-to-one. Tall boys weigh more than short boys on the average, so this is a positive correlation. But you can easily find a six-footer who weighs less than some five-footers, so the correlation is less than 1. A negative correlation is simply a statement that as one variable increases the other tends to decrease. In physics

this becomes an inverse ratio: The further you get from a light bulb the less light there is on your book; as distance increases light intensity decreases. These physical relationships often have the kindness to produce perfect correlations, but figures from business or sociology or medicine seldom work out so neatly. Even if education generally increases incomes it may easily turn out to be the financial ruination of Joe over there. Keep in mind that a correlation may be real and based on real cause and effect – and still be almost worthless in determining action in any single case.

Reams of pages of figures have been collected to show the value in dollars of a college education, and stacks of pamphlets have been published to bring these figures – and conclusions more or less based on them – to the attention of potential students. I am not quarrelling with the intention. I am in favour of education myself, particularly if it includes a course in elementary statistics. Now these figures have pretty conclusively demonstrated that people who have gone to college make more money than people who have not. The exceptions are numerous, of course, but the tendency is strong and clear.

The only thing wrong is that along with the figures and facts goes a totally unwarranted conclusion. This is the *post hoc* fallacy at its best. It says that these figures show that if *you* (your son, your daughter) attend college you will probably earn more money than if you decide to spend the next four years in some other manner. This unwarranted conclusion has for its basis the equally unwarranted assumption that since college-trained folks make more money, they make it because they went to college. Actually we don't know but that these are the people who would have made more money even if they had not gone to college. There are a couple of things that indicate rather strongly that this is so. Colleges get a disproportionate number of two groups of kids: the bright and the rich. The bright might show good earning power without college knowledge. And as for the rich ones . . . well, money breeds money in several obvious ways. Few sons of rich men are found in low-income brackets whether they go to college or not.

The following passage is taken from an article in question-and-answer form that appeared in a Sunday newspaper of enormous circulation. Maybe you will find it amusing, as I do, that the same writer once produced a piece called 'Popular Notions: True or False?'

Q: What effect does going to college have on your chances of remaining unmarried?

A: If you're a woman, it skyrockets your chances of becoming an old maid. But if you're a man, it has the opposite effect – it minimizes your chances of staying a bachelor.

Cornell University made a study of 1,500 typical middle-aged college graduates. Of the men, 93 per cent were married (compared to 83 per cent for the general population).

But of the middle-aged women graduates only 65 per cent were married. Spinsters were relatively three times as numerous among college graduates as among women of the general population.

When Susie Brown, age seventeen, reads this she learns that if she goes to college she will be less likely to get a man than if she doesn't. That is what the article says, and there are statistics from a reputable source to go with it. They go with it, but they don't back it up; and note also that while the statistics are Cornell's the conclusions are not, although a hasty reader may come away with the idea that they are.

Here again a real correlation has been used to bolster up an unproved cause-and-effect relationship. Perhaps it all works the other way around and those women would have remained unmarried even if they had not gone to college. Possibly even more would have failed to marry. If these possibilities are no better than the one the writer insists upon, they are perhaps just as valid conclusions: that is, guesses.

Indeed there is one piece of evidence suggesting that a propensity for old-maidhood may lead to going to college. Dr Kinsey seems to have found some correlation between sexuality and education, with traits perhaps being fixed at pre-college age. That makes it all the more questionable to say that going to college gets in the way of marrying.

Note to Susie Brown: It ain't necessarily so.

A medical article once pointed with great alarm to an increase in cancer among milk drinkers. Cancer, it seems, was becoming increasingly frequent in New England, Minnesota, Wisconsin, and Switzerland, where a lot of milk is produced and consumed, while remaining rare in Ceylon, where milk is scarce. For further evidence it was pointed out that cancer was less frequent in some Southern states where less milk was consumed. Also, it was pointed out, milk-drinking English women get some kinds of cancer eighteen times as frequently as Japanese women who seldom drink milk.

A little digging might uncover quite a number of ways to account for these figures, but one factor is enough by itself to show them up. Cancer is predominantly a disease that strikes in middle life or after. Switzerland and the states mentioned first are alike in having populations with relatively long spans of life. English women at the time the study was made were living an average of twelve years longer than Japanese women.

Professor Helen M. Walker has worked out an amusing illustration of the folly in assuming there must be cause and effect whenever two things vary together. In investigating the relationship between age and some physical characteristics of women, begin by measuring the angle of the feet in walking. You will find that the angle tends to be greater among older women. You might first consider whether this indicates that women grow older because they toe out, and you can see immediately that this is ridiculous. So it appears that age increases the angle between the feet, and most women must come to toe out more as they grow older.

Any such conclusion is probably false and certainly unwarranted. You could only reach it legitimately by studying the same women – or possibly equivalent groups – over a period of time. That would eliminate the factor responsible here. Which is that the older women grew up at a time when a young lady was taught to toe out in walking, while the members of the younger group were learning posture in a day when that was discouraged.

When you find somebody – usually an interested party – making a fuss about a correlation, look first of all to see if it is not one of this type, produced by the stream of events, the trend of the times. In our time it is easy to show a positive correlation between any pair of things like these: number of students in college, number of inmates in mental institutions, consumption of cigarettes, incidence of heart disease, use of X-ray machines, production of false teeth, salaries of California school teachers, profits of Nevada gambling halls. To call some one of these the cause of some other is manifestly silly. But it is done every day.

Permitting statistical treatment and the hypnotic presence of numbers and decimal points to befog causal relationships is little better than superstition. And it is often more seriously misleading. It is rather like the conviction among the people of the New Hebrides that body lice produce good health. Observation over the centuries had taught them that people in good health usually had lice and sick people very often did not. The observation itself was accurate and sound, as observations made informally over the years surprisingly often are. Not so much can be said for the conclusion to which these primitive people came from their evidence: Lice make a man healthy. Everybody should have them.

As we have already noted, scantier evidence than this – treated in the statistical mill until common sense could no longer penetrate to it – has made many a medical fortune and many a medical article in magazines, including professional ones. More sophisticated observers finally got things straightened out in the New Hebrides. As it turned out, almost everybody in those circles had lice most of the time. It was, you might say, the normal condition of man. When, however, anyone took a fever (quite possibly carried to him by those same lice) and his body became too hot for comfortable habitation, the lice left. There you have cause and effect altogether confusingly distorted, reversed, and intermingled.

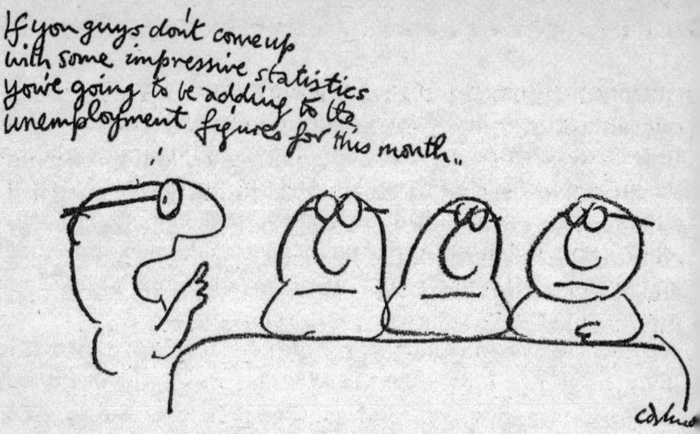

9
How to Statisticulate

Misinforming people by the use of statistical material might be called statistical manipulation; in a word (though not a very good one), statisticulation.

The title of this book and some of the things in it might seem to imply that all such operations are the product of intent to deceive. The president of a chapter of the American Statistical Association once called me down for that. Not chicanery much of the time, said he, but incompetence. There may be something in what he says,* but I am not

* Author Louis Bromfield is said to have a stock reply to critical correspondents when his mail became too heavy for individual attention. Without conceding anything and without encouraging further correspondence, it still satisfies almost everyone. The key sentence: 'There may be something in what you say.'

certain that one assumption will be less offensive to statisticians than the other. Possibly more important to keep in mind is that the distortion of statistical data and its manipulation to an end are not always the work of professional statisticians. What comes full of virtue from the statistician's desk may find itself twisted, exaggerated, over-simplified, and distorted-through-selection by salesman, public-relations expert, journalist, or advertising copywriter.

But whoever the guilty party may be in any instance, it is hard to grant him the status of blundering innocent. False charts in magazines and newspapers frequently sensationalize by exaggeration, rarely minimize anything. Those who present statistical arguments on behalf of industry are seldom found, in my experience, giving labour or the customer a better break than the facts call for, and often they give him a worse one. When has a union employed a statistical worker so incompetent that he made labour's case out weaker than it was?

As long as the errors remain one-sided, it is not easy to attribute them to bungling or accident.

One of the trickiest ways to misrepresent statistical data is by means of a map. A map introduces a fine bag of variables in which facts can be concealed and relationships distorted. My favourite trophy in this field is 'The Darkening Shadow'. It was distributed not long ago by the First National Bank of Boston and reproduced very widely – by so-called taxpayers groups, newspapers, and *Newsweek* magazine.

It reminds me of the minister who achieved great popularity among mothers in his congregation by his flattering comments on babies brought in for christening. But when the mothers compared notes not one could remember what the man had said, only that it had been 'something nice.' Turned out his invariable remark was, 'My!' (beaming) 'This *is* a baby, isn't it!'

The map shows what portion of the American income is now being taken, and spent, by the federal government. It does this by shading the areas of the states west of the Mississippi (excepting only Louisiana, Arkansas, and part of Missouri) to indicate that federal spending has become equal to the total incomes of the people of those states.

The deception lies in choosing states having large areas but, because of sparse population, relatively small incomes. With equal honesty (and equal dishonesty) the map maker might have started shading in New York or New England and come out with a vastly smaller and less impressive shadow. Using the same data he would have produced quite a different impression in the mind of anyone who looked at his map. No one would have bothered to distribute that one, though. At least, I do not know of any powerful group that is interested in making public spending appear to be smaller than it is.

If the objective of the map maker had been simply to convey information he could have done so quite easily. He could have chosen a group of in-between states whose total area bears the same relation to the area of the country that their total income does to the national income.

The thing that makes this map a particularly flagrant effort to misguide is that it is not a new trick of propaganda. It is something of a classic, or chestnut. The same bank long ago published versions of this map to show federal expenditures in 1929 and 1937, and these shortly cropped up in a standard book, *Graphic Presentation*, by Willard Cope Brinton, as horrible examples. This method 'distorts the facts', said Brinton plainly. But the First National goes right on drawing its maps, and *Newsweek* and other people who should know better – and possibly do – go right on reproducing them with neither warning nor apology.

If you think there's inflation now, consider this. At one

The darkening shadow (Western style)

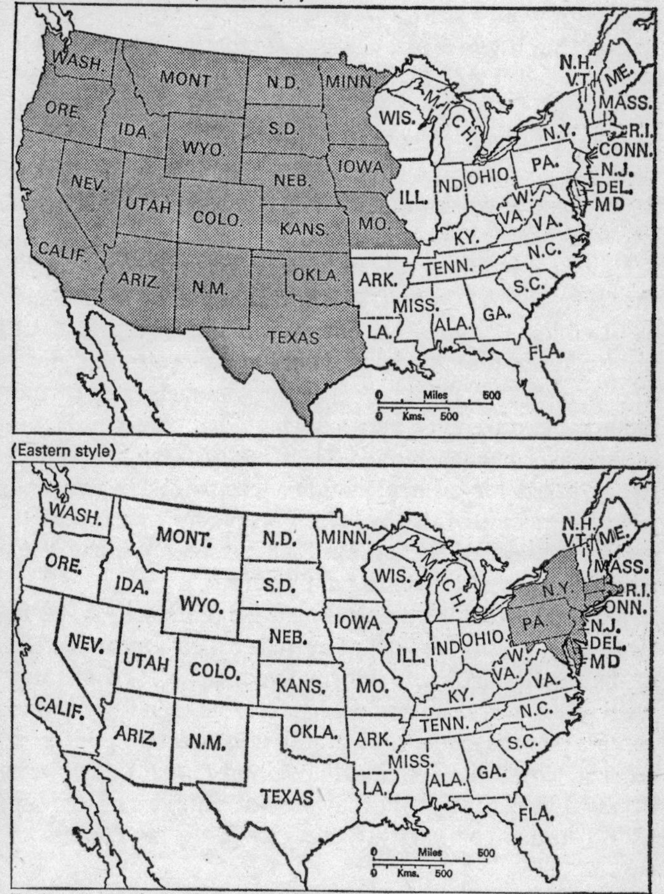

(Eastern style)

To show *we* aren't cheating, we added MD, DEL., and R.I. for good measure time the U.S. Bureau of the Census came up in its annual way with word that the 'income of the average family was $3,100'. But if you read a newspaper story on 'philanthropic giving' handed out by the Russell Sage Foundation you learned that, for the same year, it was a notable $5,004. Possibly you were pleased to learn that folks were doing so well, but you may also have been struck by how poorly that figure squared with your own observations. Possibly you know the wrong kind of people.

Now how in the world can Russell Sage and the Bureau of the Census be so far apart? The Bureau is talking in medians, as of course it should be, but even if the Sage people are using a mean the difference should not be quite this great. The Russell Sage Foundation, it turns out, discovered this remarkable prosperity by producing what can only be described as a phoney family. Their method, they explained (when asked for an explanation), was to divide the total personal income of the American people by 149,000,000 to get an average of $1,251 for each person. 'Which,' they added, 'becomes $5,004 in a family of four.'

This odd piece of statistical manipulation exaggerates in two ways. It uses the kind of average called a mean instead of the smaller and more informative median ... something we worked over in an earlier chapter. And then it goes on to assume that the income of a family is in direct proportion to its size. Now I have four children, and I wish things were disposed in that way, but they are not. Families of four are by no means commonly twice as wealthy as families of two.

In fairness to the Russell Sage statisticians, who may be presumed innocent of desire to deceive, it should be said that they were primarily interested in making a picture of giving rather than of getting. The funny figure for family incomes was just a by-product. But it spread its deception no less effectively for that, and it remains a prime example of why little faith can be placed in an unqualified statement of average.

For a spurious air of precision that will lend all kinds of weight to the most disreputable statistic, consider the decimal. Ask a hundred citizens how many hours they slept last

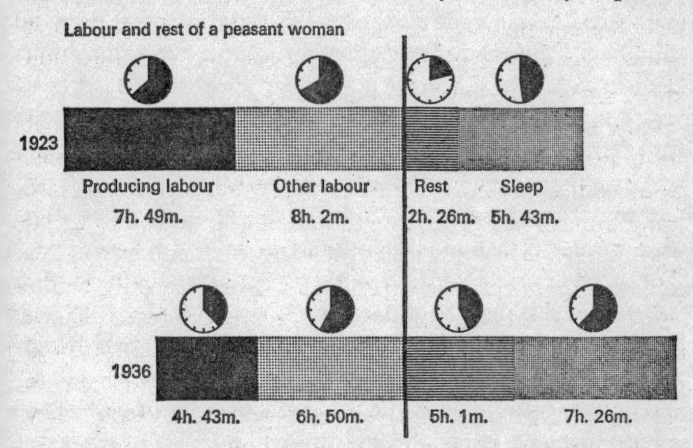

Chart adapted from U.S.S.R. (Scientific Publishing Institute of Pictorial Statistics).

night. Come out with a total of, say, 783·1. Any such data are far from precise to begin with. Most people will miss their guess by fifteen minutes or more, and there is no assurance that the errors will balance out. We all know someone who will recall five sleepless minutes as half a night of tossing insomnia. But go ahead, do your arithmetic, and announce that people sleep an average of 7·831 hours a night. You will sound as if you knew precisely what you were talking about. If you had been so foolish as to declare only that people sleep 7·8 (or 'almost 8') hours a night, there would have been nothing striking about it. It would have sounded like what it was, a poor approximation and no more instructive than almost anybody's guess.

Karl Marx was not above achieving a spurious air of precision in the same fashion. In figuring the 'rate of surplus-value' in a mill he began with a splendid collection of assumptions, guesses, and round numbers: 'We assume the waste to be 6 per cent . . . the raw material . . . costs in round numbers £342. The 10,000 spindles . . . cost, we will assume, £1 per spindle. . . . The wear and tear we put at 10 per cent. . . . The rent of the building we suppose to be £300 . . .' He says, 'the above data, which may be relied upon, were given me by a Manchester spinner.'

From these approximations Marx calculates that: 'The rate of surplus-value is therefore $\frac{80}{52} = 153\frac{11}{13}$ per cent.' For a ten-hour day this gives him 'necessary labour $= 3\frac{31}{33}$ hours and surplus-labour $= 6\frac{2}{33}$.'

There's a nice feeling of exactness to that two thirty-thirds of an hour, but it's all bluff.

Percentages offer a fertile field for confusion. And like the ever-impressive decimal they can lend an aura of precision to the inexact. The United States Department of Labor's *Monthly Labor Review* once stated that of the offers of part-time household employment with provisions for car fare, in Washington, D.C., during a specified month, 4·9 per cent were at $18 a week. This percentage, it turned out, was based on precisely two cases, there having been only forty-one offers altogether. Any percentage figure based on a small number of cases is likely to be misleading. It is more informative to give the figure itself. And when the percentage is carried out to decimal places you begin to run the scale from the silly to the fraudulent.

'Buy your Christmast presents now and save 100 per cent', advises an advertisement. This sounds like an offer worthy of old Santa himself, but it turns out to be merely a confusion of base. The reduction is only fifty per cent. The saving is one hundred per cent of the reduced or new price, it is true, but that isn't what the offer says.

Likewise when the president of a flower growers' association said, in a newspaper interview, that 'flowers are 100 per cent cheaper than four months ago,' he didn't mean that florists were now giving them away. But that's what he said.

In her *History of the Standard Oil Company*, Ida M. Tarbell went even further. She said that 'price cutting in the southwest ... ranged from 14 to 220 per cent'. That would call for seller paying buyer a considerable sum to haul the oily stuff away.

The Columbus *Dispatch* declared that a manufactured product was selling at a profit of 3,800 per cent, basing this on a cost of $1·75 and a selling price of $40. In calculating percentage of profits you have a choice of methods (and you are obligated to indicate which you are using). If figured on cost, this one comes to a profit of 2,185 per cent; on selling price, 95·6 per cent. The *Dispatch* apparently used a method of its own and, as so often seems to happen, got an exaggerated figure to report.

Even the *New York Times* lost the Battle of the Shifting Base in publishing an Associated Press story from Indianapolis:

> The depression took a stiff wallop on the chin here today. Plumbers, plasterers, carpenters, painters and others affiliated with the Indianapolis Building Trades Unions were given a 5 per cent increase in wages. That gave back to the men one-fourth of the 20 per cent cut they took last winter.

Sounds reasonable on the face of it – but the decrease has been figured on one base – the pay the men were getting in the first place – while the increase uses a smaller base, the pay level after the cut.

You can check on this bit of statistical misfiguring by supposing, for simplicity, that the original wage was $1 an hour. Cut twenty per cent, it is down to 80 cents. A five per cent increase on that is 4 cents, which is not one-fourth but one-fifth of the cut. Like so many presumably honest mistakes, this one somehow managed to come out an exaggeration which made a better story.

All this illustrates why to offset a pay cut of fifty per cent you must get a raise of one hundred per cent.

It was the *Times* also that once reported that, for a fiscal year, air mail 'lost through fire was 4,863 pounds, or a percentage of but 0·00063'. The story said that planes had carried 7,715,741 pounds of mail during the year. An insurance company basing its rates in that way could get into a pack of trouble. Figure the loss and you'll find that it came to 0·063 per cent or one hundred times as great as the newspaper had it.

It is the illusion of the shifting base that accounts for the trickiness of adding discounts. When a hardware jobber offers '50 per cent and 20 per cent off list', he doesn't mean a seventy per cent discount. The cut is sixty per cent since the twenty per cent is figured on the smaller base left after taking off fifty per cent.

A good deal of bumbling and chicanery have come from adding together things that don't add up but merely seem to. Children for generations have been using a form of this device to prove that they don't go to school.

You probably recall it. Starting with 365 days to the year you can subtract 122 for the one-third of the time you spend in bed and another 45 for the three hours a day used in eating. From the remaining 198 take away 90 for summer vacation and 21 for Christmas and Easter vacations. The days that remain are not even enough to provide for Saturdays and Sundays.

Too ancient and obvious a trick to use in serious business, you might say. But the United Automobile Workers insist in their monthly magazine, *Ammunition*, that it is still being used against them.

The wide, blue yonder lie also turns up during every strike. Every time there is a strike, the Chamber of Commerce advertises that the strike is costing so many millions of dollars a day.

They get the figure by adding up all the cars that would have been made if the strikers had worked full time. They add in losses to suppliers in the same way. Everything possible is added in, including street car fares and the loss to merchants in sales.

The similar and equally odd notion that percentages can be added together as freely as apples has been used against authors. See how convincing this one, from *The New York Times Book Review*, sounds.

The gap between advancing book prices and authors' earnings, it appears, is due to substantially higher production and material costs. Item: plant and manufacturing expenses alone have risen as much as 10 to 12 per cent over the last decade, materials are up 6 to 9 per cent, selling and advertising expenses have climbed upwards of 10 per cent. Combined boosts add up to a minimum of 33 per cent (for one company) and to nearly 40 per cent for some of the smaller houses.

Actually, if each item making up the cost of publishing this book has risen around ten per cent, the total cost must have climbed by about that proportion also. The logic that permits adding those percentage rises together could lead to all sorts of flights of fancy. Buy twenty things today and find that each has gone up five per cent over last year. That 'adds up' to one hundred per cent, and the cost of living has doubled. Nonsense.

It's all a little like the tale of the roadside merchant who was asked to explain how he could sell rabbit sandwiches so cheap. 'Well,' he said, 'I have to put in some horse meat too. But I mix 'em fifty-fifty: one horse, one rabbit.'

A union publication used a cartoon to object to another variety of unwarranted adding-up. It showed the boss adding one regular hour at $1·50 to one overtime hour at $2·25 to one double-time hour at $3 for an average hourly wage of $2·25. It would be hard to find an instance of an average with less meaning.

Another fertile field for being fooled lies in the confusion between percentage and percentage points. If your profits should climb from three per cent on investment one year to six per cent the next, you can make it sound quite modest by calling it a rise of three percentage points. With equal validity, you can describe it as a one hundred per cent increase. For loose handling of this confusing pair watch particularly the public-opinion pollers.

Percentiles are deceptive too. When you are told how Johnny stands compared to his classmates in algebra or some aptitude, the figure may be a percentile. It means his rank in each one hundred students. In a class of three hundred, for instance, the top three will be at the 99 percentile, the next three at the 98, and so on. The odd thing about percentiles is that a student with a 99-percentile rating is probably quite a bit superior to one standing at 90, while those at the 40 and 60 percentiles may be of nearly equal achievement. This comes from the habit that so many characteristics have of clustering about their own average, forming the 'normal' bell curve we mentioned in an early chapter.

Occasionally a battle of the statisticians develops, and even the most unsophisticated observer cannot fail to smell a rat. Honest men get a break when statisticulators fall out. The Steel Industry Board has pointed out some of the monkey business in which both steel companies and unions have indulged. To show how good business had been in the year just ended (as evidence that the companies could well afford a raise), the union compared that year's productivity with the productivity of 1939 – a year of especially low volume. The companies, not to be outdone in the deception derby, insisted on making their comparisons on a basis of money received by the employees rather than average hourly earnings. The point to this was that so many workers had been on part time in the earlier year that their incomes were bound to have grown even if wage rates had not risen at all.

Time magazine, notable for the consistent excellence of its graphics, published a chart that is an amusing example of how statistics can pull out of the bag almost anything that may be wanted. Faced with a choice of methods, equally valid, one favouring the management viewpoint and the other favouring labour, *Time* simply used both. The chart was really two charts, one superimposed upon the other. They used the same data.

One showed wages and profits in billions of dollars. It was evident that although both were rising, the increase in wages in the last year was roughly twice that in profits. And that wages involved perhaps six times as many dollars as profits did. The great inflationary pressure, it appeared, came from wages.

The other part of the dual chart expressed the changes as percentages of increase. The wage line was relatively flat. The profit line shot sharply upwards. Profits, it might be inferred, were principally responsible for inflation.

You could take your choice of conclusions. Or, perhaps better, you could easily see that neither element could properly be singled out as the guilty one. It is sometimes a substantial service simply to point out that a subject in

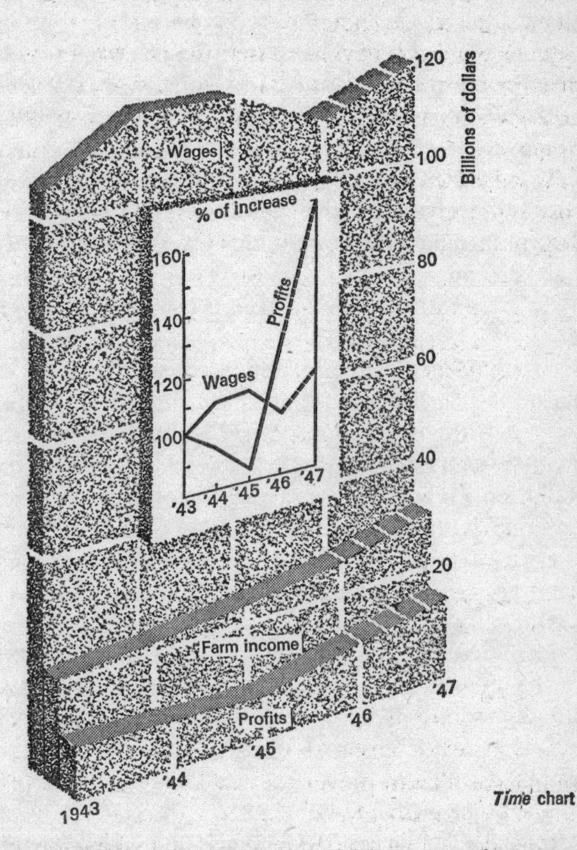

Redrawn with the kind permission of Time *magazine as an example of a non-lying chart.*

controversy is not as open-and-shut as it has been made to seem.

Index numbers are vital matters to millions of people now that wage rates are often tied to them. It is perhaps worth noting what can be done to make them dance to any man's music.

To take the simplest possible example, let's say that milk cost 10p a quart last year and bread was 10p a loaf. This year milk is down to 5p and bread is up to 20p. Now what

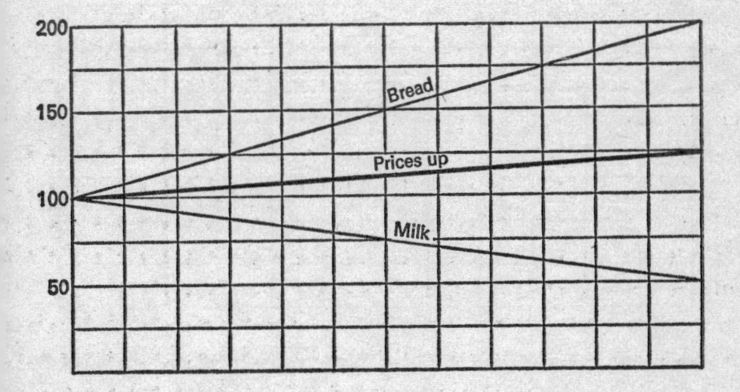

would you like to prove? Cost of living up? Cost of living down? Or no change?

Consider last year as the base period, making the prices of that time 100 per cent. Since the price of milk has since dropped to half (50 per cent) and the price of bread has doubled (200 per cent) and the average of 50 and 200 is 125, prices have gone up 25 per cent.

Try it again, taking this year as base period. Milk used to cost 200 per cent as much as it does now and bread was selling for 50 per cent as much. Average: 125 per cent. Prices used to be 25 per cent higher than they are now.

To prove that the cost level hasn't changed at all we simply switch to the geometric average and use either period as the base. This is a little different from the arithmetic average, or mean, that we have been using but it is a perfectly legitimate kind of figure and in some cases the

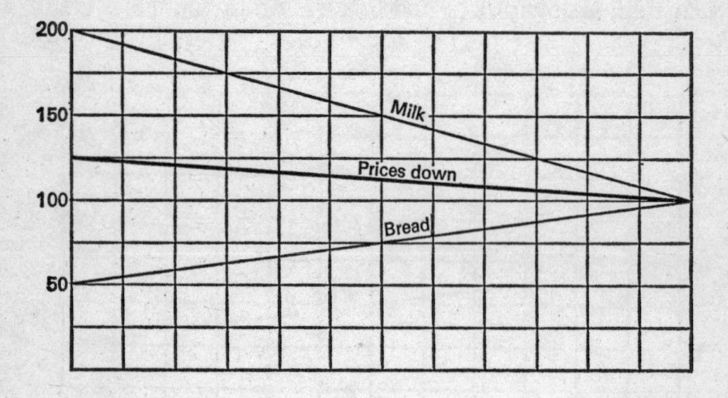

most useful and revealing. To get the geometric average of three numbers you multiply them together and derive the cube root. For four items, the fourth root; for two, the square root. Like that.

Take last year as the base and call its price level 100. Actually you multiply the 100 per cent for each item together and take the root, which is 100. For this year, milk being at 50 per cent of last year and bread at 200 per cent, multiply 50 by 200 to get 10,000. The square root, which is the geometric average, is 100. Prices have not gone up *or* down.

The fact is that, despite its mathematical base, statistics is as much an art as it is a science. A great many manipulations and even distortions are possible within the bounds of propriety. Often the statistician must choose among methods, a subjective process, and find the one that he will use to represent the facts. In commercial practice he is about as unlikely to select an unfavourable method as a copywriter is to call his sponsor's product flimsy and cheap when he might as well say light and economical.

Even the man in academic work may have a bias (possibly unconscious) to favour, a point to prove, an axe to grind.

This suggests giving statistical material, the facts and figures in newspapers and books, magazines and advertising, a very sharp second look before accepting any of them. Sometimes a careful squint will sharpen the focus. But arbitrarily rejecting statistical methods makes no sense either. That is like refusing to read because writers sometimes use words to hide facts and relationships rather than to reveal them. After all, a political candidate in Florida not long ago made considerable capital by accusing his opponent of 'practising celibacy'. A New York exhibitor of the motion picture *Quo Vadis* used huge type to quote the *New York Times* as calling it 'historical pretentiousness'. And the makers of Crazy Water Crystals, a proprietary medicine, have been advertising their product as providing 'quick, ephemeral relief'.

10 How to Talk Back to a Statistic

So far, I have been addressing you rather as if you were a pirate with a yen for instruction in the finer points of cutlass work. In this concluding chapter I'll drop that literary device. I'll face up to the serious purpose that I like to think lurks just beneath the surface of this book: explaining how to look a phoney statistic in the eye and face it down; and no less important, how to recognize sound and usable data in that wilderness of fraud to which the previous chapters have been largely devoted.

Not all the statistical information that you may come upon can be tested with the sureness of chemical analysis or of what goes on in an assayer's laboratory. But you can prod the stuff with five simple questions, and by finding the answers avoid learning a remarkable lot that isn't so.

Who Says So?

About the first thing to look for is bias – the laboratory with something to prove for the sake of a theory, a reputation, or a fee; the newspaper whose aim is a good story; labour or management with a wage level at stake.

Look for conscious bias. The method may be direct misstatement or it may be ambiguous statement that serves as well and cannot be convicted. It may be selection of favourable data and suppression of unfavourable. Units of measurement may be shifted, as with the practice of using one year for one comparison and sliding over to a more favourable year for another. An improper measure may be used: a mean where a median would be more informative (perhaps all too informative), with the trickery covered by the unqualified word 'average'.

Look sharply for unconscious bias. It is often more dangerous. In the charts and predictions of many statisticians and economists in 1928 it operated to produce remarkable things. The cracks in the economic structure were joyously overlooked, and all sorts of evidence were adduced and statistically supported to show that we had no more than entered the stream of prosperity.

It may take at least a second look to find out who-says-so. The who may be hidden by what Stephen Potter, the *Lifemanship* man, would probably call the 'O.K. name'. Anything smacking of the medical profession is an O.K. name. Scientific laboratories have O.K. names. So do universities, more especially ones eminent in technical work. The writer who proved a few chapters back that higher education jeopardizes a girl's chance to marry made good use of the O.K. name of Cornell. Please note that while the data came from Cornell, the conclusions were entirely the writer's own. But the O.K. name helps you carry away a misimpression of 'Cornell University says . . .'

When an O.K. name is cited, make sure that the authority stands behind the information, not merely somewhere alongside it.

You may have read a proud announcement by the Chicago *Journal of Commerce*. That publication had made a survey. Of 169 corporations that replied to a poll on price gouging and hoarding, two-thirds declared that they were absorbing price increases produced by the police action, or undeclared war, in which the United States was then as usual engaged in the Far East. 'The survey shows,' said the *Journal* (look sharp whenever you meet those words!), 'that corporations have done exactly the opposite of what the enemies of the American business system have charged.' This is an obvious place to ask, 'Who says so?' since the *Journal of Commerce* might be regarded as an interested party. It is also a splendid place to ask our second test question:

How Does He Know?

It turns out that the *Journal* had begun by sending its questionnaires to 1,200 large companies. Only fourteen per cent had replied. Eighty-six per cent had not cared to say anything in public on whether they were hoarding or price gouging.

The *Journal* had put a remarkably good face on things, but the fact remains that there was little to brag about. It came down to this: Of 1,200 companies polled, nine per cent said they had not raised prices, five per cent said they had, and eighty-six per cent wouldn't say. Those that had replied constituted a sample in which bias might be suspected.

Watch out for evidence of a biased sample, one that has been selected improperly or – as with this one – has selected itself. Ask the question we dealt with in an early chapter: Is the sample large enough to permit any reliable conclusion?

Similarly with a reported correlation: Is it big enough to mean anything? Are there enough cases to add up to any significance? You cannot, as a casual reader, apply tests of significance or come to exact conclusions as to the adequacy of a sample. On a good many of the things you see reported, however, you will be able to tell at a glance – a good long glance, perhaps – that there just weren't enough cases to convince any reasoning person of anything.

What's Missing?

You won't always be told how many cases. The absence of such a figure, particularly when the source is an interested one, is enough to throw suspicion on the whole thing. Similarly a correlation given without a measure of reliability (probable error, standard error) is not to be taken very seriously.

Watch out for an average, variety unspecified, in any matter where mean and median might be expected to differ substantially.

Many figures lose meaning because a comparison is missing. An article in *Look* magazine says, in connection with Mongolism, that 'one study shows that in 2,800 cases, over half of the mothers were 35 or over'. Getting any meaning from this depends upon your knowing something about the ages at which women in general produce babies. Few of us know things like that.

Here is a pollution note from nearly a generation ago, from a *New Yorker* magazine 'Letter from London':

> The Ministry of Health's recently published figures showing that in the week of the great fog the death rate for Greater London jumped by twenty-eight hundred were a shock to the public, which is used to regarding Britain's unpleasant climatic effects as nuisances rather than as killers . . . The extraordinary lethal properties of this winter's prize visitation . . .

But how lethal *was* the visitation? Was it exceptional for the death rate to be that much higher than usual in a week? All such things do vary. And what about ensuing weeks? Did the death rate drop below average, indicating that if the fog killed people they were largely those who would have died shortly anyway? The figure sounds impressive, but the absence of other figures takes away most of its meaning.

Sometimes it is percentages that are given and raw figures that are missing, and this can be deceptive too. Long ago, when Johns Hopkins University had just begun to admit women students, someone not particularly enamoured of co-education reported a real shocker: Thirty-three and one-third per cent of the women at Hopkins had married faculty members! The raw figures gave a clearer picture. There were three women enrolled at the time, and one of them had married a faculty man.

Some years ago the Boston Chamber of Commerce chose its American Women of Achievement. Of the sixteen among them who were also in *Who's Who*, it was announced that they had 'sixty academic degrees and eighteen children'. That sounds like an informative picture of the group until you discover that among the women were Dean Virginia Gildersleeve and Mrs Lillian M. Gilbreth. Those two had a full third of the degrees between them. And Mrs Gilbreth, about whose offspring *Cheaper by the Dozen* was written, supplied two-thirds of the children.

A corporation was able to announce that its stock was held by 3,003 persons, who had an average of 660 shares each. This was true. It was also true that of the two million shares of stock in the corporation three men held three-quarters and three thousand persons held the other one-fourth among them.

If you are handed an index, you may ask what's missing there. It may be the base, a base chosen to give a distorted picture. A national labour organization once showed that indexes of profits and production had risen much more rapidly after the depression than an index of wages had. As an argument for wage increases this demonstration lost its potency when someone dug out the missing figures. It could be seen then that profits had been almost bound to rise more rapidly in percentage than wages simply because profits had reached a lower point, giving a smaller base.

Sometimes what is missing is the factor that caused a change to occur. This omission leaves the implication that some other, more desired, factor is responsible. Figures published one year attempted to show that business was on the upgrade by pointing out that April retail sales were greater than in the year before. What was missing was the fact that Easter had come in March in the earlier year and in April in the later year.

A report of a great increase in deaths from cancer in the last quarter-century is misleading unless you know how much of it is a product of such extraneous factors as these: Cancer is often listed now where 'causes unknown' was formerly used; autopsies are more frequent, giving surer diagnoses; reporting and compiling of medical statistics are more complete; and people more frequently reach the most susceptible ages now. And if you are looking at total deaths rather than the death rate, don't neglect the fact that there are more people now than there used to be.

Did Somebody Change the Subject?

When assaying a statistic, watch out for a switch somewhere between the raw figure and the conclusion. One thing is all too often reported as another.

As just indicated, more reported cases of a disease are not always the same thing as more cases of the disease. A straw-vote victory for a candidate is not always negotiable at the polls. An expressed preference by a 'cross-section' of a magazine's readers for articles on world affairs is no final proof that they would read the articles if they were published.

A year came when encephalitis cases reported in the central valley of California were triple the figure for the worst previous year. Many alarmed residents shipped their children away. But when the reckoning was in, there had been no great increase in deaths from sleeping sickness. What had happened was that state and federal health people had come in in great numbers to tackle a long-time problem; as a result of their efforts a great many low-grade cases were recorded that in other years would have been overlooked, possibly not even recognized.

It is all reminiscent of the way that Lincoln Steffens and Jacob A. Riis, as New York newspapermen, once created a crime wave. Crime cases in the papers reached such proportions, both in numbers and in space and big type given to them, that the public demanded action. Theodore Roosevelt, as president of the reform Police Board, was seriously embarrassed. He put an end to the crime wave simply by asking Steffens and Riis to lay off. It had all come about simply because the reporters, led by those two, had got into competition as to who could dig up the most burglaries and whatnot. The official police record showed no increase at all.

'The British male over 5 years of age soaks himself in a hot tub on an average of 1·7 times a week in the winter and 2·1 times in the summer,' says a newspaper story. 'British women average 1·5 baths a week in the winter and 2·0 in the summer.' The source is a Ministry of Works hot-water survey of '6,000 representative British homes'. The sample was representative, it says, and seems quite adequate in size to justify the conclusion in the San Francisco *Chronicle*'s amusing headline: BRITISH HE'S BATHE MORE THAN SHE'S.

The figures would be more informative if there were some indication of whether they are means or medians. However, the major weakness is that the subject has been changed. What the Ministry really found out is how often these people said they bathed, not how often they did so. When a subject is as intimate as this one is, with the British bath-taking tradition involved, saying and doing may not be the same thing at all. British he's may or may not bathe oftener than she's; all that can safely be concluded is that they say they do.

Here are some more varieties of change-of-subject to watch out for.

A back-to-the-farm movement was discerned when a census showed half a million more farms in the U.S. than five years earlier. But the two counts were not talking about the same thing. The definition of farm used by the Bureau of the Census had been changed; it took in at least 300,000 farms that would not have been so listed under the earlier definition.

Strange things crop up when figures are based on what people say – even about things that seem to be objective facts. Census reports have shown more people at thirty-five years of age, for instance, than at either thirty-four or thirty-six. The false picture comes from one family member's reporting the ages of the others and, not being sure of the exact ages, tending to round them off to a familiar multiple of five. One way to get around this: ask for birth dates instead.

The 'population' of a large area in China was 28 million. Five years later it was 105 million. Very little of that increase was real; the great difference could be explained only by taking into account the purposes of the two enumerations and the way people would be inclined to feel about being counted in each instance. The first census was for tax and military purposes, the second for famine relief.

Something of the same sort has happened in the United States. A decennial census found more people in the sixty-five-to-seventy age group than there were in the fifty-five-to-sixty group ten years before. The difference could not be accounted for by immigration. Most of it could be a product of large-scale falsifying of ages by people eager to collect social security. Also possible is that some of the earlier ages were understated out of vanity.

Another kind of change-of-subject is represented by U.S. Senator William Langer's cry, back in days when San Francisco's notorious island was a boarding house for hard cases and hotels charged less than they do now, that 'we could take a prisoner from Alcatraz and board him at the Waldorf-Astoria cheaper . . .' The North Dakotan was referring to earlier statements that it cost eight dollars a day to maintain a prisoner at Alcatraz, 'the cost of a room at a good San Francisco hotel'. The subject has been changed from total maintenance cost (Alcatraz) to hotel-room rent alone.

The *post hoc* variety of pretentious nonsense is another way of changing the subject without seeming to. The change of something *with* something else is presented as *because of*.

The magazine *Electrical World* once offered a composite chart in an editiorial on 'What Electricity Means to America'. You could see from it that as 'electrical horsepower in factories' climbed, so did 'average wages per hour'. At the same time 'average hours per week' dropped. All these things are long-time trends, of course, and there is no evidence at all that any one of them has produced any other.

And then there are the firsters. Almost anybody can claim to be first in *something* if he is not too particular what it is. At the end of 1952 two New York newspapers were each insisting on first rank in grocery advertising. Both were right too, in a way. The *World-Telegram* went on to explain that it was first in full-run advertising, the kind that appears in all copies, which is the only kind it runs. The *Journal-American* insisted that total linage was what counted and that it was first in that. This is the kind of reaching for a superlative that leads the weather reporter on the radio to label a quite normal day 'the hottest June second since 1967'.

Change-of-subject makes it difficult to compare cost when you contemplate borrowing money either directly or in the form of instalment buying. Six per cent sounds like six per cent – but it may not be at all.

If you borrow £100 from a bank at six per cent interest and pay it back in equal monthly instalments for a year, the price you pay for the use of the money is about £3. But another six per cent loan, on the basis sometimes called £6 on the £100, will cost you twice as much. That's the way most automobile loans are figured. It is very tricky.

The point is that you don't have the £100 for a year. By the end of six months you have paid back half of it. If you are charged at £6 on the £100, or six per cent of the amount, you really pay interest at nearly twelve per cent.

Even worse was what happened to some careless purchasers of freezer-food plans in America. They were quoted a figure of anywhere from six to twelve per cent. It sounded like interest, but it was not. It was an on-the-dollar figure and, worst of all, the time was often six months rather than a year. Now £12 on the £100 for money to be paid back regularly over half a year works out to something like forty-eight per cent real interest. It is no wonder that so many customers defaulted and so many food plans blew up.

Sometimes the semantic approach will be used to change the subject. Here is an item from *Business Week* magazine.

> Accountants have decided that 'surplus' is a nasty word. They propose eliminating it from corporate balance sheets. The Committee on Accounting Procedure of the American Institute of Accountants says: ... Use such descriptive terms as 'retained earnings' or 'appreciation of fixed assets'.

This one is from a newspaper story reporting Standard Oil's record-breaking revenue and net profit of a million dollars a day.

> Possibly the directors may be thinking some time of splitting the stock for there may be an advantage ... if the profits per share do not look so large....

Does It Make Sense?

'Does it make sense?' will often cut a statistic down to size when the whole rigmarole is based on an unproved assumption. You may be familiar with the Rudolf Flesch readability formula. It purports to measure how easy a piece of prose is to read, by such simple and objective items as length of words and sentences. Like all devices for reducing the imponderable to a number and substituting arithmetic for judgement, it is an appealing idea. At least it has appealed to people who employ writers, such as newspaper publishers, even if not to many writers themselves. The assumption in the formula is that such things as word length determine readability. This, to be ornery about it, remains to be proved.

A man named Robert A. Dufour put the Flesch formula to trial on some literature that he found handy. It showed 'The Legend of Sleepy Hollow' to be half again as hard to read as Plato's *Republic*. The Sinclair Lewis novel *Cass Timberlane* was rated more difficult than an essay by Jacques Maritain, 'The Spiritual Value of Art'. A likely story.

Many a statistic is false on its face. It gets by only because the magic of numbers brings about a suspension of common sense. Leonard Engel, in a *Harper's* article, has listed a few of the medical variety.

> An example is the calculation of a well-known urologist that there are eight million cases of cancer of the prostate gland in the United States – which would be enough to provide 1·1 carcinomatous prostate glands for every male in the susceptible age group! Another is a prominent neurologist's estimate that one American in twelve suffers from migraine; since migraine is responsible for a third of chronic headache cases, this would mean that a quarter of us must suffer from disabling headaches. Still another is the figure of 250,000 often given for the number of multiple sclerosis cases; death data indicate that there can be, happily, no more than thirty to forty thousand cases of this paralytic disease in the country.

Hearings on amendments to the Social Security Act have been haunted by various forms of a statement that makes sense only when not looked at closely. It is an argument that goes like this: Since life expectancy is only about sixty-three years, it is a sham and a fraud to set up a social-security plan with a retirement age of sixty-five, because virtually everybody dies before that.

You can rebut that one by looking around at people you know. The basic fallacy, however, is that the figure refers to expectancy at birth, and so about half the babies born can expect to live longer than that. The figure, incidentally, came from the 1939–41 period and remained in use long after it was out of date. Presumably the current figure, calculated a generation later, of 69·7 will produce a new and equally silly argument to the effect that practically everybody now lives to be sixty-five.

Product planning at a big electrical-appliance company was going great guns some years ago on the basis of a declining birth rate, something that had been taken for granted for a long time. Plans called for emphasis on small-capacity appliances, apartment-size refrigerators. Then one of the planners had an attack of common sense: He came out of his graphs and charts long enough to notice that he and his co-workers and his friends and his neighbours and his former classmates with few exceptions either had three or four children or planned to. This led to some open-minded investigating and charting – and the company shortly turned its emphasis most profitably to big-family models. It is to be hoped that the planners will respond more quickly to the present turnaround.

The impressively precise figure is something else that contradicts common sense. A study reported in New York City newspapers announced that a working woman living with her family needed a weekly pay cheque of $40·13 for adequate support. Anyone who has not suspended all logical processes while reading his paper will realize that the cost of keeping body and soul together cannot be calculated to the last cent. But there is a dreadful temptation; '$40·13' sounds so much more knowing than 'about $40'.

You are entitled to look with the same suspicion on the report, some years ago, by the American Petroleum Industries Committee that the average yearly tax bill for automobiles is $51·13.

Extrapolations are useful, particularly in that form of soothsaying called forecasting trends. But in looking at the figures or the charts made from them, it is necessary to remember one thing constantly: The trend-to-now may be a fact, but the future trend represents no more than an educated guess. Implicit in it is 'everything else being equal' and 'present trends continuing'. And somehow everything else refuses to remain equal, else life would be dull indeed.

For a sample of the nonsense inherent in uncontrolled extrapolation, consider the trend of television. The number of sets in American homes increased around 10,000 per cent in one five-year period early in the game. Project this for the next five years and you'd have found that there were about to be a couple of thousand million of the things, Heaven forfend, or forty sets to the family. If you want to be sillier yet, begin with a base year even earlier in the telly scheme of things and you can just as well 'prove' that each family will soon have not forty but forty thousand sets.

What Harry Truman did to Tom Dewey, in a U.S. Presidential-election upset unparalleled before or since, is nothing to what Truman did to the poll people. A Government research man, Morris Hansen, has called that Gallup election forecast 'the most publicized statistical error in human history'.

It was a paragon of accuracy, however, compared to some of our most widely used estimates of future population, which have earned a nationwide horselaugh. As late as 1938 a Presidential commission loaded with experts doubted that the population of the United States would ever reach 140 million; it was 12 million more than that just twelve years later. Yet college textbooks published so recently that they were still in use at that time predicted a peak population of not more than 150 million and figured it would take until about 1980 to reach it. These fearful underestimates came from assuming that a trend would continue without change. A similar assumption a century ago did as badly in the opposite direction because it assumed continuation of population-increase rate of 1790 to 1860. In his second message to Congress, Abraham Lincoln predicted the U.S. population would reach 251,689,914 in 1930.

Not long after that, in 1874, Mark Twain summed up the nonsense side of extrapolation in *Life on the Mississippi:*

> In the space of one hundred and seventy-six years the Lower Mississippi has shortened itself two hundred and forty-two miles. That is an average of a trifle over one mile and a third per year. Therefore, any calm person, who is not blind or idiotic, can see that in the Old Oölitic Silurian period, just a million years ago next November, the Lower Mississippi River was upward of one million three hundred thousand miles long, and stuck out over the Gulf of Mexico like a fishing-rod. And by the same token any person can see that seven hundred and forty-two years from now the Lower Mississippi will be only a mile and three-quarters long, and Cairo and New Orleans will have joined their streets together, and be plodding comfortably along under a single mayor and a mutual board of aldermen. There is something fascinating about science. One gets such wholesale returns of conjecture out of such a trifling investment of fact.

More About Penguins and Pelicans

For further information about books available from Penguins please write to Dept EP, Penguin Books Ltd, Harmondsworth, Middlesex UB7 0DA.

In the U.S.A.: For a complete list of books available from Penguins in the United States write to Dept CS, Penguin Books, 625 Madison Avenue, New York, New York 10022.

In Canada: For a complete list of books available from Penguins in Canada write to Penguin Books Canada Ltd, 2801 John Street, Markham, Ontario L3R 1B4.

In Australia: For a complete list of books available from Penguins in Australia write to the Marketing Department, Penguin Books Australia Ltd, P.O. Box 257, Ringwood, Victoria 3134.

Facts from Figures

M. J. Moroney

The enormous success and rapid expansion of statistical techniques in recent years is ample proof of the need for them. The reader will find here a comprehensive introduction to the possibilities of the subject; he is given the how and the why and the wherefore by which he can recognize the kind of problem where Statistics pays dividends. Common sense and simple arithmetic will carry the reader through this book. Every symbol, every principle is explained and illustrated with examples drawn from a wide variety of subjects. The author writes from experience, for he knows the limitations to the usefulness of statistical technique, and appreciates the difficulties of the non-mathematician. The book ranges from purely descriptive statistics, through probability theory, the game of Crown and Anchor, the design of sampling schemes, production quality control, correlation and ranking methods, to the analysis of variance and co-variance.

Use and Abuse of Statistics

W. J. Reichmann

Without statistics much of the machinery of modern life would grind slowly to a standstill. The subject which is said to have prompted Disraeli to quip: 'There are three sorts of lie: lies, damned lies, and statistics', is nevertheless an essential tool in science, industry, commerce, and other fields.

This book introduces the reader in readily comprehensible terms to the world of averages, probabilities, percentages, indexes, samples, and trends. Explaining the meaning of such expressions, the author shows how and for what purpose statistics may usefully be employed and how they should certainly not be employed. The result is an introductory discussion of a fascinating application of mathematics.

'The really important part of the book – and at this level it is the best I have read on the subject – is his discussion of what statistics is, and does . . . this book will prove a boon' – *Financial Times*